Sierra Challenge:
The Construction of the Chihuahua al Pacífico Railroad

news articles and photographs
by Glenn Burgess

compiled and edited
by Don Burgess

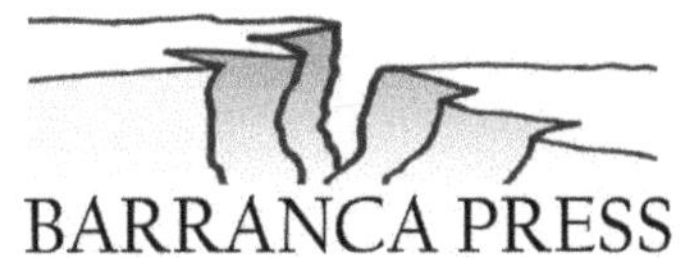

BARRANCA PRESS

Designed by LMB Noudéhou.
www.barrancapress.com

FIRST EDITION, June 2013
Chapter headings and photo captions clarified December 2014.

The articles from the *El Paso Times* and the *Fort Worth Star-Telegram* are used with the appropriate permission.

The article from the *CF&I Blast* is used courtesy of the Bessemer Historical Society/CF&I Archives.

Photos by Glenn Burgess are identified by G.B., while photos by Don Burgess are identified by D.B.. Captions for photos as they originally appeared in a newspaper begin with bold capitalized phrases.

The Owen map is from the Special Collections Research Center, California State University, Fresno.

The Cuiteco photo by photographer Ira Kneeland is used by permission of the Special Collections Research Center, California State University, Fresno.

The SCOP map is from Glenn Burgess' personal archives.

Translation of Spanish texts is by Don Burgess.

The front cover photograph of fastening rails is by Glenn Burgess.

The back cover photograph of the Autovía is by Don Burgess.

The cover design is by Sarah Markes.

ISBN 978-1-939604-22-4

Library of Congress Control Number: 2013903968

For worldwide distribution.
Printed in the United States of America.

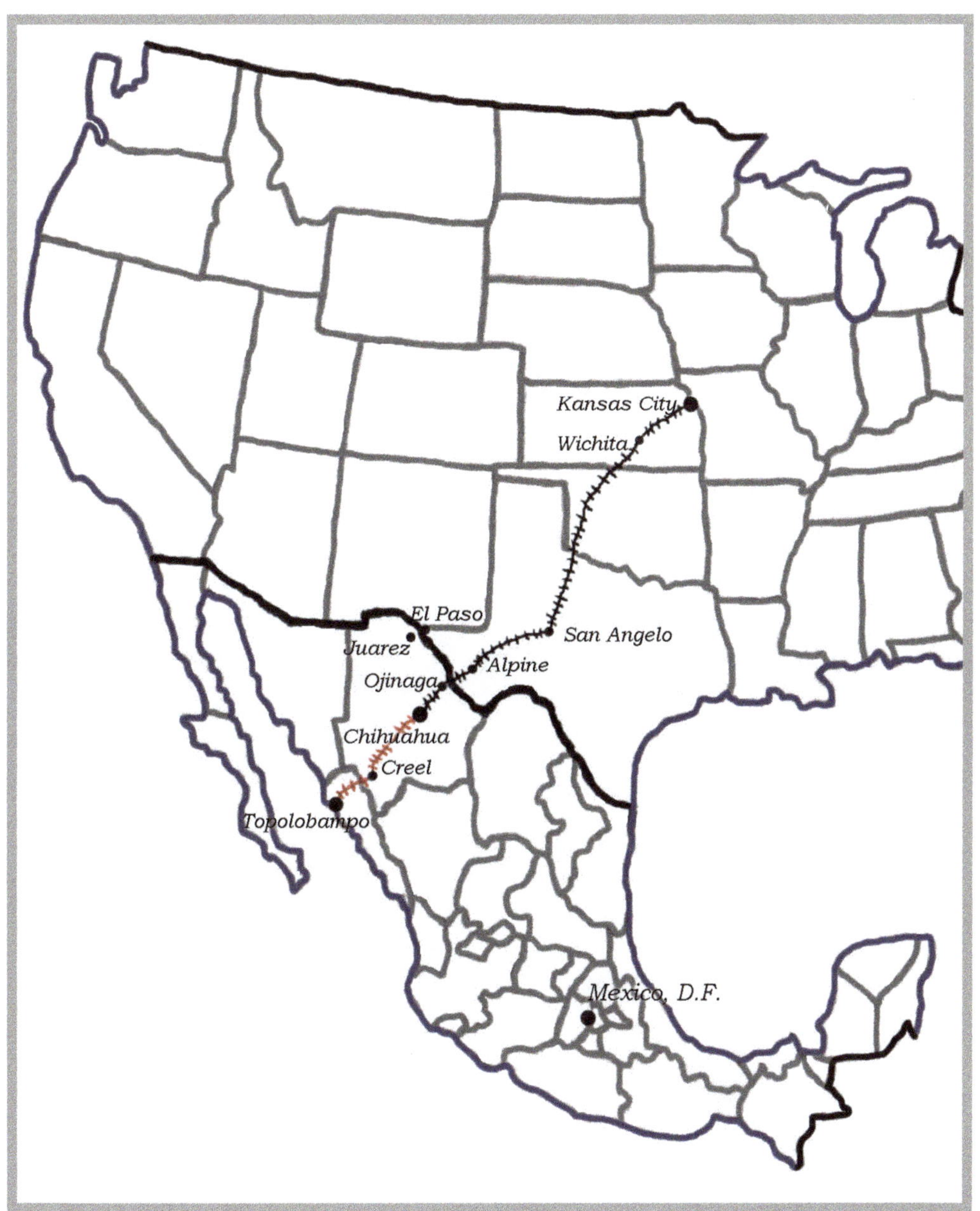

1.1. Sketch showing Arthur E. Stilwell's idea connecting Kansas City to the Pacific Ocean by rail. The section west of Chihuahua City is now known as the Chihuahua al Pacífico Railroad (CHEPE).

Erle Stanley Gardner described the challenge of building a railroad across the canyons of Chihuahua and Sinaloa in this way:

> You stand there on the brink of that huge canyon, look down into the depths, and think that anyone who expected a railroad to go through that country would have cheerfully entertained a scheme to make a pogo stick large enough to enable its user to jump to the moon.
>
> It is almost exactly as though the railroad to the Grand Canyon which starts at Williams, Arizona, was brought right up to the south rim of the canyon with the expectation of the engineers to take trains down to the Colorado River below and then follow the canyon to Yuma.

Erle Stanley Gardner
Neighborhood Frontiers (p.248)

Dedication

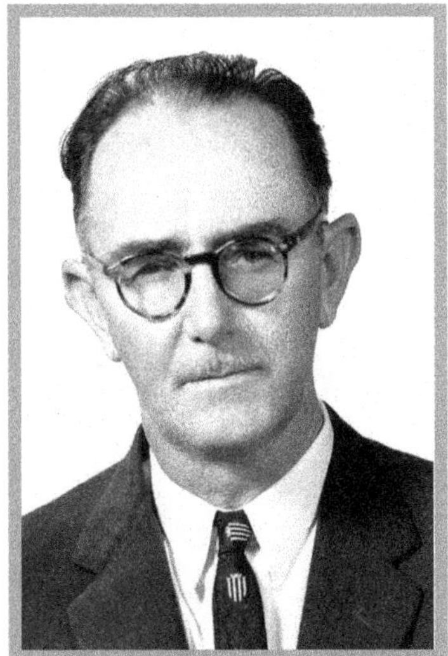

Ing. Francisco M. Togno (1907-1991)

In accordance with my dad's wish (and I am in complete agreement), this book is dedicated to Ing. Francisco M. Togno[1]. As Dad wrote:

> One man stands out as the reason for completing the railroad. Ing. Francisco Togno started as a local contractor near Creel, then as the engineer in charge of locating the final route, the engineer in charge of the construction office at Chihuahua City, later as chief of construction of all railroads in Mexico, and now raised to the position of assistant minister of public works.
>
> It is this engineer who convinced three presidents of Mexico that the project was feasible and that it should be completed. He believed that the original promoter Stilwell[2] used sound economic reasoning in proposing and initiating the railroad, and dedicated his life work to the fulfillment of the Stilwell dream.

One of the laborers who worked with him said to me: "Era un hombre muy fino" (He was a very fine man.), and I do not know anyone who would disagree.

1 Even though the spelling of this name is Togno, the engineers that I know pronounce it as Toño, which is the Italian pronunciation (Italian gn = ñ), but most people in Mexico pronounce it Togno.
The photographer is unknown.

2 Actually there were other promoters before Stilwell.

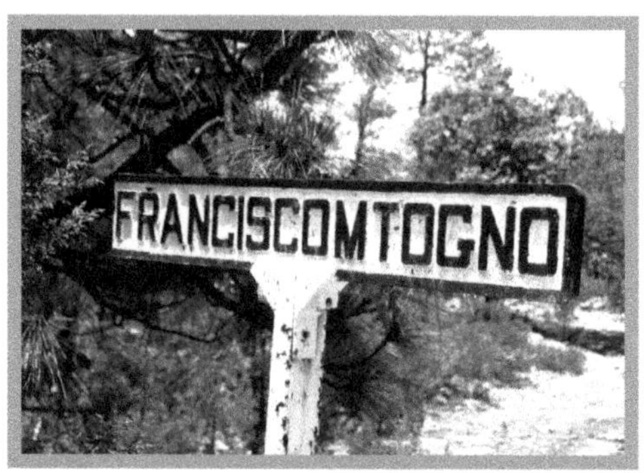

1.3. Railroad siding named for Ing. Francisco M. Togno, between El Divisadero and Areponápuchi (photo D.B. 2012).

Contents

Acknowledgements

My thanks to the ones who read the manuscript at various stages— Lisa María Burgess, Luís Urías, Kay Burgess, Joe Burgess, and William Merrill. Bryan Wilson and Joe Burgess made adjustments on some of the photographs. The work of editing and formatting the book was done by Barranca Press.

And a special thanks to the railroad workers and engineers who shared with me their knowledge and excitement about the railroad. The more I am around them, the more I realize what a magnificent engineering feat the construction of this railroad was. I cannot begin to do it justice in this book.

Introduction

by Don Burgess

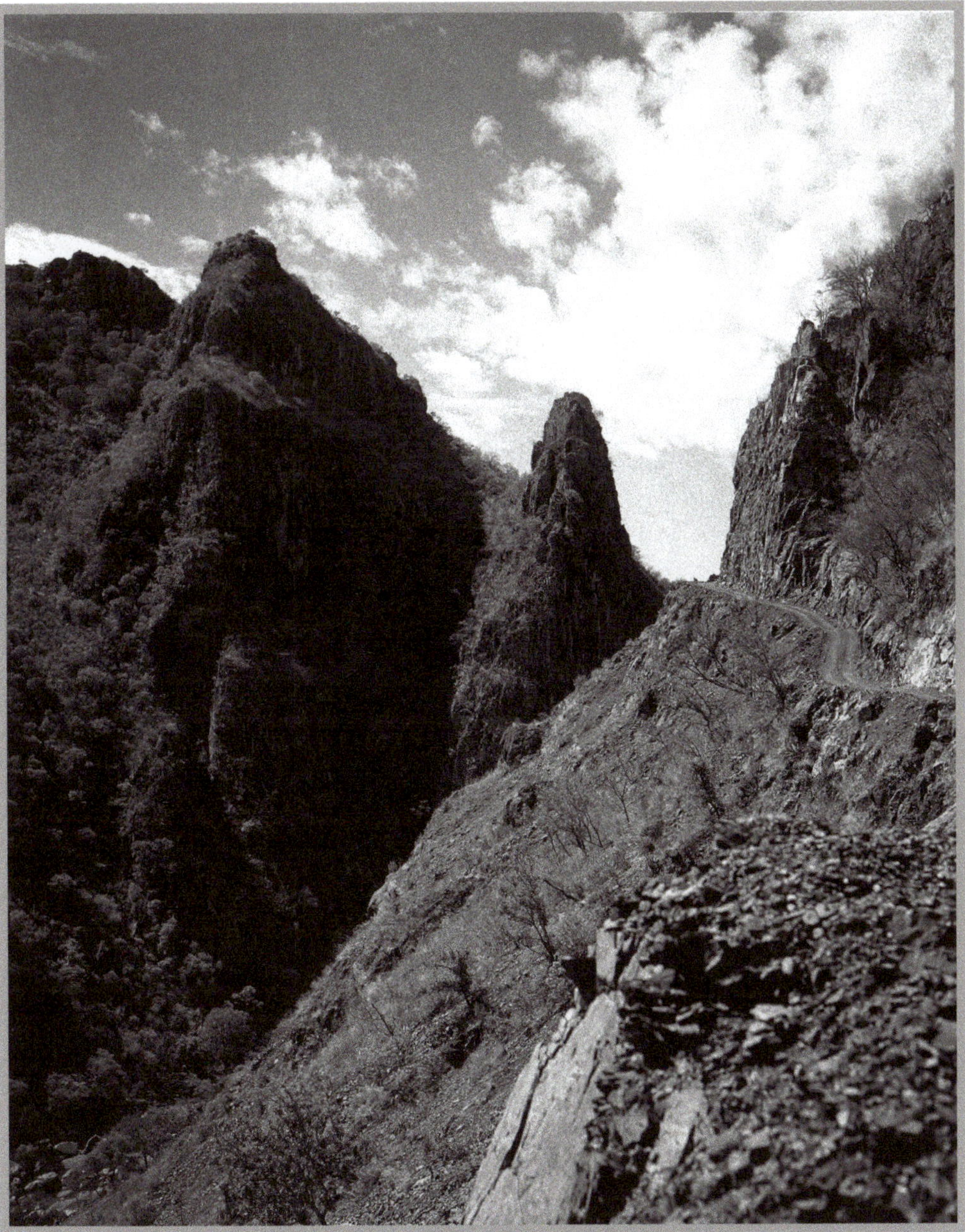

2.1. A rugged pass built to reach the railroad construction site near Témoris (photo G.B. 1955).

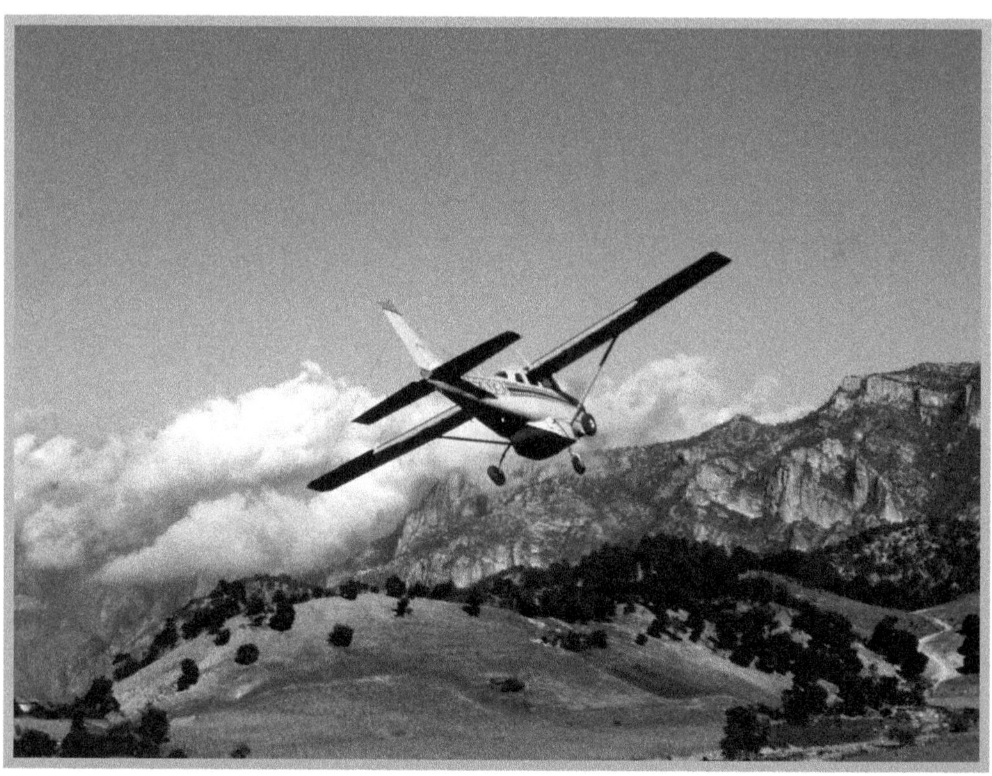

2.2. Plane leaving Rocoroibo, Chihuahua. The Chínipas River Canyon can be seen to the left (photo D.B. c.1995).

On October 13, 1995, a small plane circled the log cabin where my wife and I lived in a remote area of the Sierra Tarahumara of Chihuahua, Mexico. As I ran to a clearing in the woods, the pilot slowed his speed and dropped a plastic bag tied with colored ribbons. Inside were a small rock and a note. The bag missed the clearing and landed in the forest. Several Tarahumaras and I searched for the bag, and when it was found, I tore it open and read the note: "Your father is not expected to live through the day. If you want me to fly you out, I will wait at Rocoroibo [the closest airstrip]. If you understand this message, walk in a circle waving a towel."

Not having a towel handy, I pulled out my handkerchief and began waving it as I walked in a circle, feeling rather self-conscious as a number of Tarahumaras watched my strange antics.

As the plane flew off, I consulted with my wife, Marie. She was a nurse, and about twenty Tarahumaras were waiting for medical attention from her and a doctor who had come down from Arizona. Some of the sick people had walked five hours just to get there. Marie could not leave.

So, I jumped in a pick-up, along with a couple of friends, and headed up the mountain towards where the plane would be waiting, an hour and fifteen minute drive away. Part way up the mountain, we were met by a friend who lived near the airstrip, had radio contact with the pilot, and knew the situation. He had come on his motorcycle to pick me up. I transferred to the back of the motorcycle, and the two men who had come with me drove the pick-up back to our house.

Within ten minutes of arriving at the airstrip, the plane was racing down the steep, short, narrow airstrip, which goes over a hill and has a curve in it. Normally my heart skips a few beats when we take off from that strip, but this time my mind was on other things. Soon we were skimming the mountain ridges and then, as if the bottom had dropped out of the mountains, we were over the Oteros River Canyon (which becomes the Chínipas River Canyon). From the top of Huichúachi Mountain on our right, to the river below us, is a drop of about 7,000 feet.

Sitting in the plane and looking out at the mountains transversed by the Chihuahua al Pacífico Railroad, I began to think about my dad. Back in 1955, he had introduced me to this rugged landscape and the impossible dream that was the railroad.

My dad had been part of the generation which saw more changes in transportation than perhaps any other. When he was two years old (1907), he was taken in a covered wagon from Coleman, Texas to the family's new dry-land farm near Lubbock, Texas. And he had seen the age of super-sonic planes and rockets taking people into space.

Burgess Family History in Chihuahua

According to research by my brother Jack, the Burgess interest in the Chihuahua City, Ojinaga, Presidio and Alpine parts of the area covered by the railroad goes back to 1846 when John D. Burgess (We think he was a relative.) of the Missouri Volunteers went with Doniphan's troops into Mexico and fought in the Battle of Sacramento just north of Chihuahua City. He and several of his friends liked Mexico so much they decided to stay, and they married Mexican women. It is rumored that they made money, along with other "gringos", by taking Apache scalps, for which the Chihuahua government paid quite well. Then he and Ben Leaton, Milton Faber and John W. Spencer moved to Presidio, Texas (where the Chihuahua al Pacífico Railroad now crosses into the United States). Faber ranched and Leaton farmed and established Fort Leaton. Burgess and Spencer lived as traders, packing supplies to Fort Davis and Fort Stockton. On one of those trips Spencer picked up a rock to throw at an ornery burro and noticed something different about the rock. He later found out that it was high grade silver ore and thus the silver mine of Shafter was discovered. Burgess became famous when he held off some Comanche warriors for several days at Alpine at what was called for many years Burgess Springs and is presently called Kokernot Springs.

When most of the railroad articles were written, he lived in Alpine with his wife, Merlyn, and three boys: Jack, Don, and Joe. At that time he operated a camera and curios shop in Alpine, wrote feature articles for the *El Paso Times* and *Fort Worth Star-Telegram*, and taught journalism and photography at Sul Ross College. I remember one of his articles about a Volkswagen that flipped over on the highway after swerving to miss a Jackrabbit. Dad visited the driver in the hospital, and the article, entitled something like: "New Yorker has encounter with Texas size Jackrabbit and comes in second best," made it to the Associated Press and was published across the country.

It was in 1955 that he was assigned by the *Times* and the *Star-Telegram* to cover the final construction of the Chihuahua al Pacífico Railroad, the old Kansas City, Mexico & Orient. I was sixteen at the time, and, when he invited me to go with him, I jumped at the opportunity, quitting my summer job working on a survey crew in the Big Bend National Park.

For the next two weeks, the Mexican government showed us the construction, our guide being Jorge Togno, the brother of the engineer most responsible for the completion of the railroad, Francisco Togno. We not only saw the railroad, we also visited the then working copper mine located in the spectacular canyon of

La Bufa, and the Huites Dam project in Sinaloa which was completed, after forty years, in 1995, the highest dam in Latin America.

Dad made several other trips to visit the railroad construction and, in 1959, between my junior and senior year at Texas Western College in El Paso, I had the privilege of working on the construction. Our friendship with the Togno family made such a job possible. I worked in the mountain village of Cuiteco as a rod-man with the engineers, surveying the tunnels and bridges in that area. My pay was about $1.70 (17 pesos) per day, enough to pay room and board, and save 10 dollars after six weeks. I worked under Ing. Felipe Anzaldua, a pleasant man who I remember trying to teach me that Mexican music was more than "Cielito Lindo," "La Cucaracha," and a few other Mexican songs of which gringos sometimes know a few strains. I can still sing a line or two of one song he taught me: "Y si quiere saber de mi pasado, es preciso decir una mentira...." I also played baseball

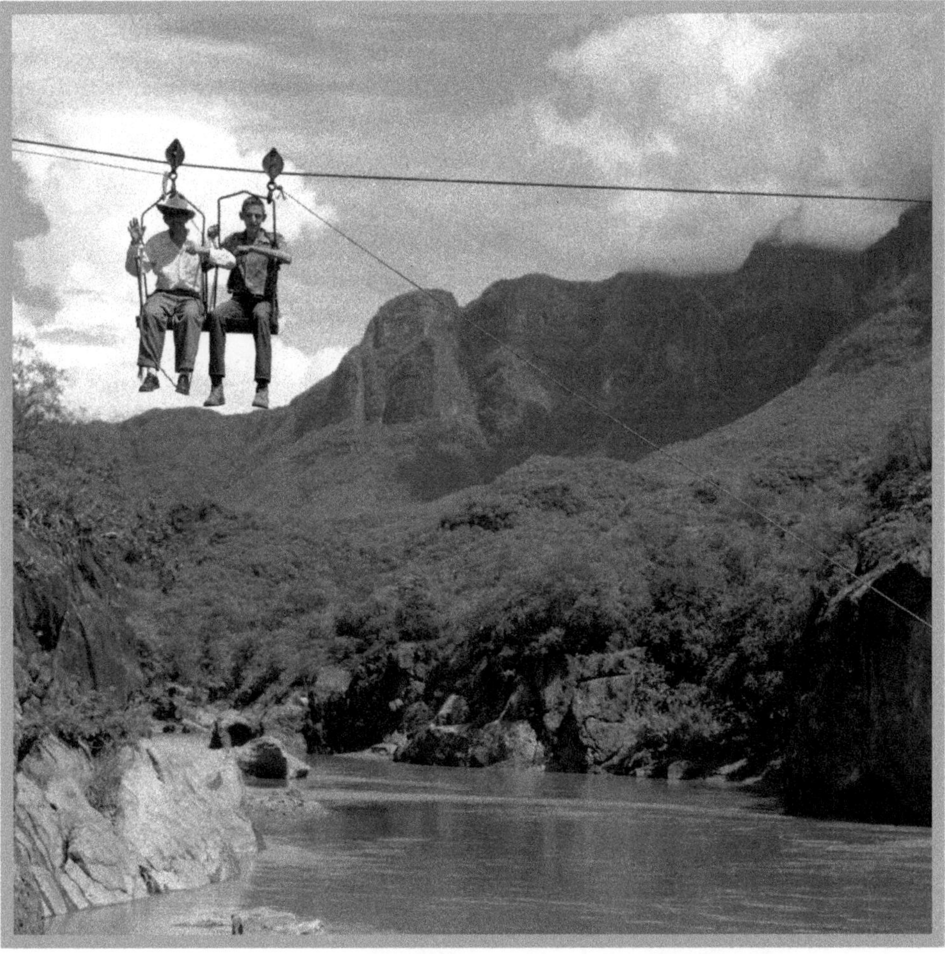

2.3. Jorge Togno and Don Burgess crossing the Chínipas River near the bridge in 1955 (photo G.B.).

Week-end activities for construction engineers and workers included baseball, basketball, and soccer.

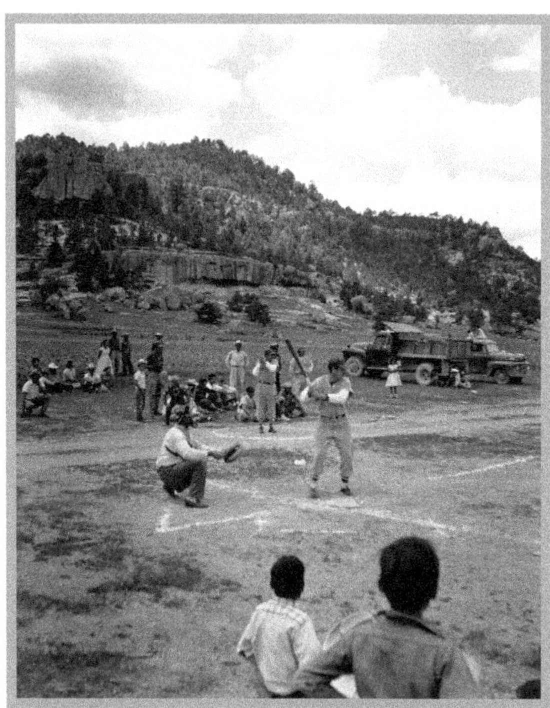

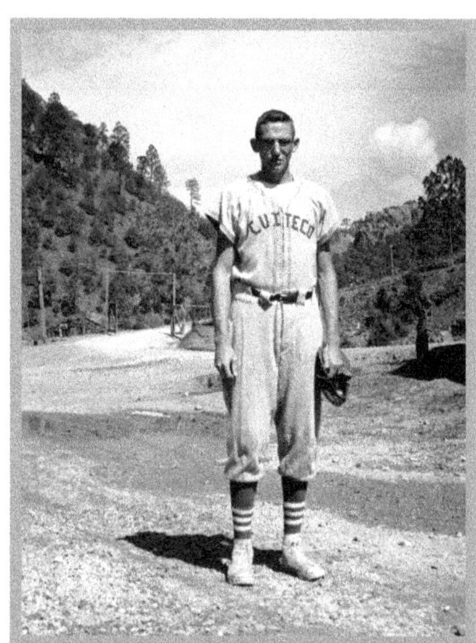

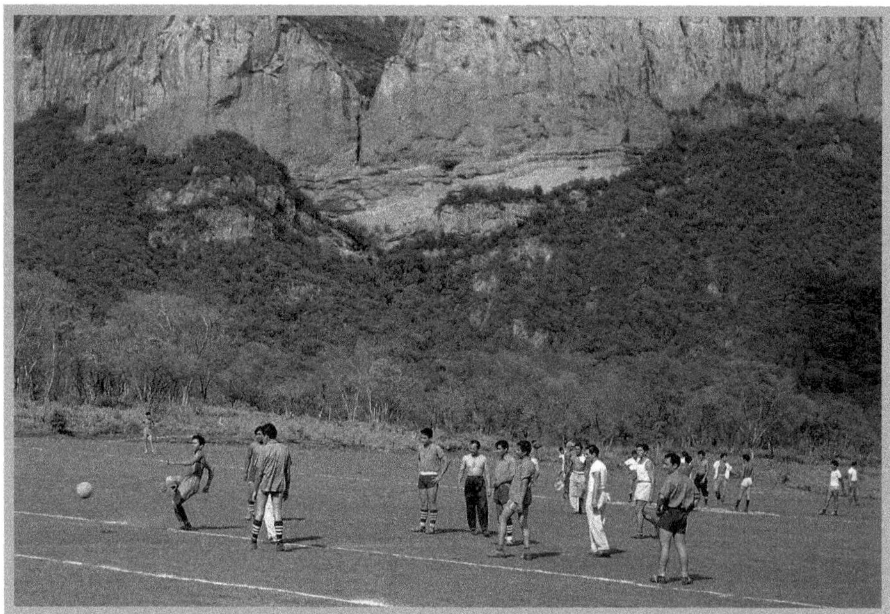

2.4. Playing baseball in Creel in 1959 (photo D.B.). **2.5.** Don was in uniform in Cuiteco the same year (photographer unknown 1959). **2.6.** Playing soccer at the construction camp near the Huites Canyon in Sinaloa (photo G.B. 1955).

and basketball for the local railroad team and on week-ends we would travel to other construction camps and towns for games.

During that summer, I enjoyed the country and the Tarahumara[1] people (who call themselves Ralámuli) so much that I returned to college wondering, "What excuse can I find for returning to the Sierra Tarahumara to live?" Then I realized that, with the opening of the railroad, the Tarahumara people would have to face "civilization" in ways they had never had to do before, and without books in their language to help them preserve their language and culture, and books to help them face a new way of life, the chances were high they would lose much of what they had to offer the world. At that time, only a few primers and parts of the New Testament had been prepared in the Tarahumara Alta[2] and nothing in the Tarahumara Baja. These are the two major Tarahumara dialects.

So, I finished college and pursued a Master's degree in history, the title of my thesis being "History of Missionary Efforts Among the Tarahumara Since 1604." Then I went to the University of Oklahoma to study descriptive linguistics. After that, I returned to the Sierra Tarahumara and began to study the language and to help the Tarahumara people translate and write books. As my father often said, "As a result of that first trip to visit the railroad, more than fifty articles and books have been written about the Sierra Tarahumara." These have included newspaper articles, two master's thesis, linguistic and anthropological studies, bilingual Tarahumara-Spanish books written by Tarahumaras concerning their culture (the latest is on Tarahumara uses of corn), a translation of the New Testament as well as parts of the Old Testament in the Baja Tarahumara language, and the preparation of medical books on tuberculosis, diarrhea, and breastfeeding in the native language.

Dad celebrated his 90th birthday in August of 1995 and I was encouraging him to put his articles and photos into a book on the construction of the railroad, but it never happened. When the plane landed in Chihuahua City, I phoned home and was given the news that Dad had already died, actually about two hours before the pilot had dropped his message to me. The pilot graciously took me on to Safford, Arizona where I attended the funeral service.

1 In the writings of the early catholic missionaries, the forms tarahumar and tarahumares were used for singular and plural, and the form tarahumara as a qualifier for persons and objects. Today most people in the Sierra only use the words tarahumara and tarahumaras.

2 The Jesuits David Brambila and Carlos Infante Díaz had respectively translated Mark and Luke. And Ken Hilton of the Wycliffe Bible Translators was working in Samachique on the New Testament, which was published in 1972.

So, the task of doing something with Dad's materials fell to me. Data on the period prior to 1900 needed to be added, and something needed to be said about how the railroad had affected the local cultures. Dad especially wanted to make sure that Ing. Francisco Togno received the credit he deserved for his part in the construction, and wanted to dedicate the book to him. I have mostly used Dad's articles as they were published, and I have added notes throughout the book which I hope will give a more complete picture and will be of interest to the reader.[3] Three of the articles were so similar to others that I did not include them. In the back I have included two interviews with railroad workers who were personal friends of mine, one a Tarahumara man and the other a man of Japanese descent. And finally there is a short discussion of how the railroad has affected the Tarahumara and "Mestizo" cultures in the Sierra.

The real jewels of this publication are the black and white photographs taken by Dad, done with a 4 x 5 Speed Graphic camera which he had purchased in Santa Fe, NM in 1949. He also took a 16 mm movie of our first trip, but he sent the only copy to Mexico City many years ago, and we were never able to locate it. I have added a few photos of my own, as well as some taken by other people.

Two of the best books written about the railroad are John Leeds Kerr and Frank Donavan's *Destination Topolobampo*, and Francisco Almada's *El Ferrocarril de Chihuahua al Pacífico*. The strength of the first book lies in the coverage of the construction on the United States side. The strength of the second lies in the coverage of the histories of the various railroad projects in Chihuahua. Other books that need to be mentioned are: *Memoria de la Construcción del Ferrocarril Chihuahua al Pacífico*, David M. Pletcher's *Rails, Mines, and Progress: Seven American Promoters in Mexico, 1867-1911*, and L.L. Waters' *Steel Trails to Santa Fe*. A documentary-type film, which made use of some of Dad's photos and where I was interviewed, was produced by Rebo Studios for Japanese and American public TV and was called *Train Ride To The Sky* (1994).

3 I have also corrected minor misspellings and added accent marks.

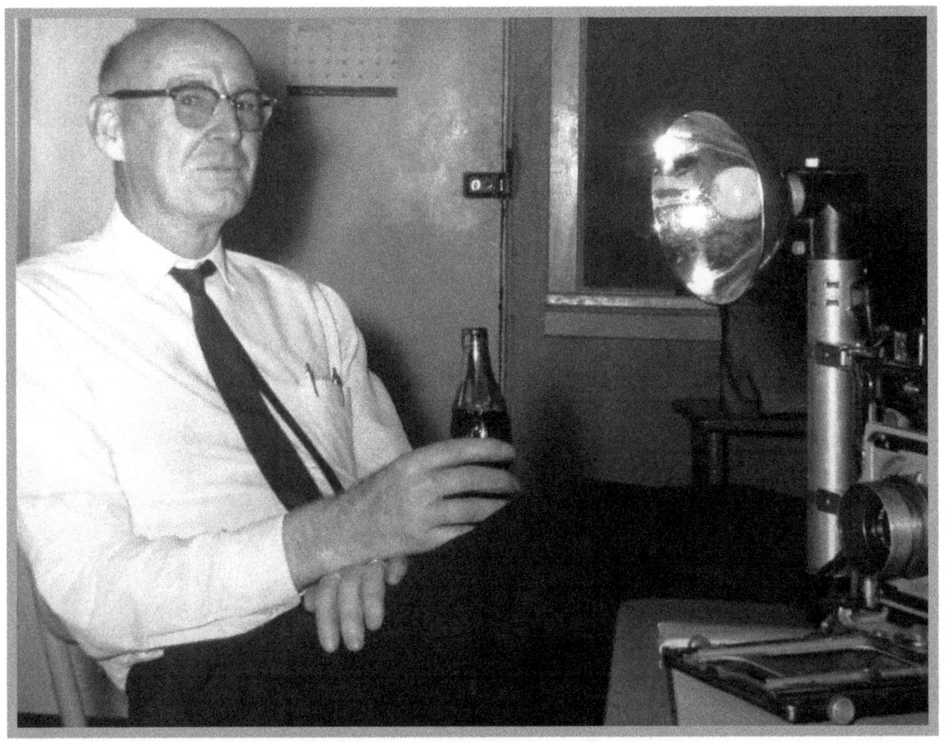

2.7. Glenn Burgess with his Speed Graphic camera, taken in the 1960s in Alpine, Texas (photographer unknown).

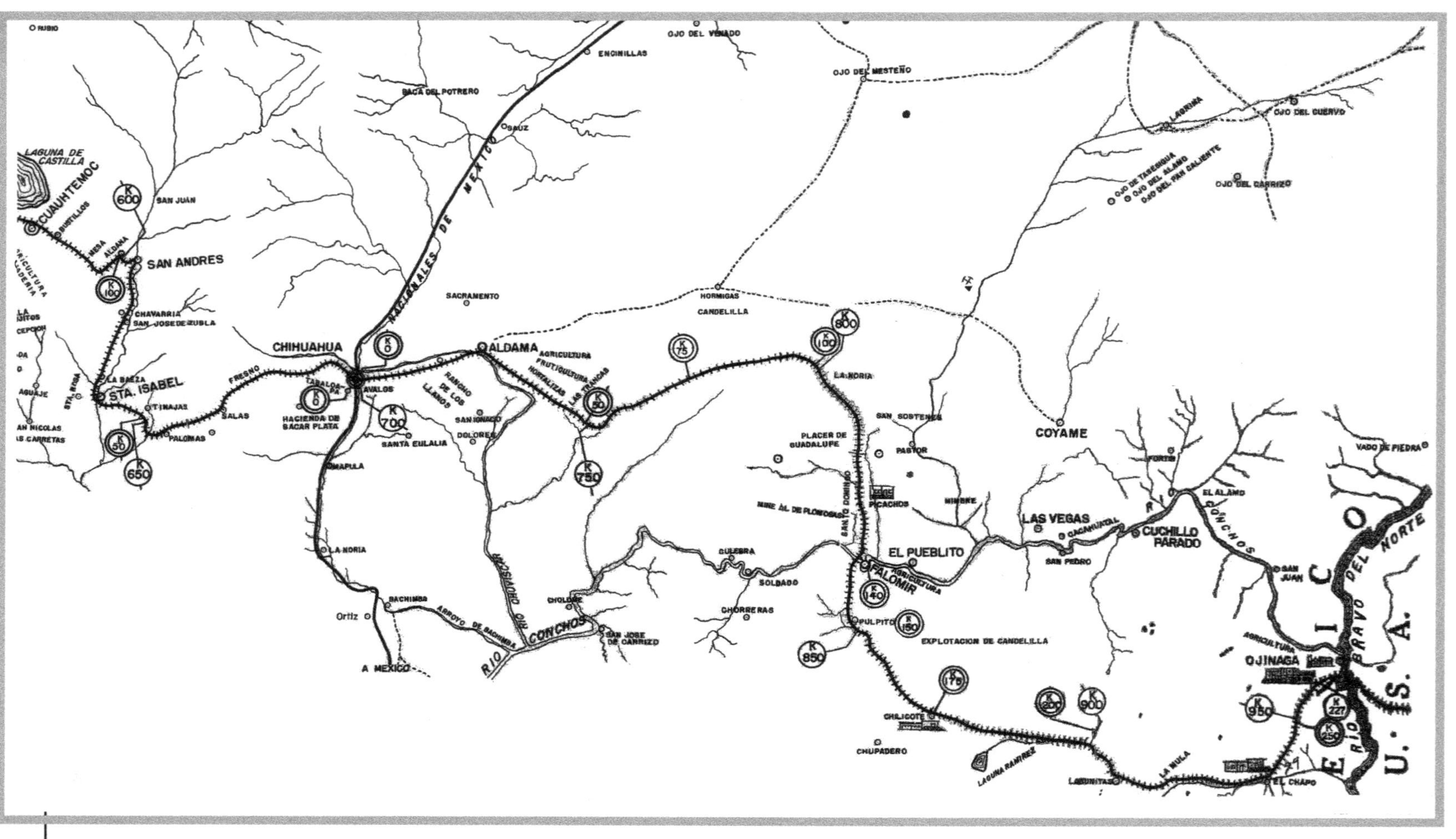

2.8. Beginning of the line: Map from Mexico's Department of Communications and Public Works, showing the lines that come into Chihuahua City, Mexico (SCOP 1953).

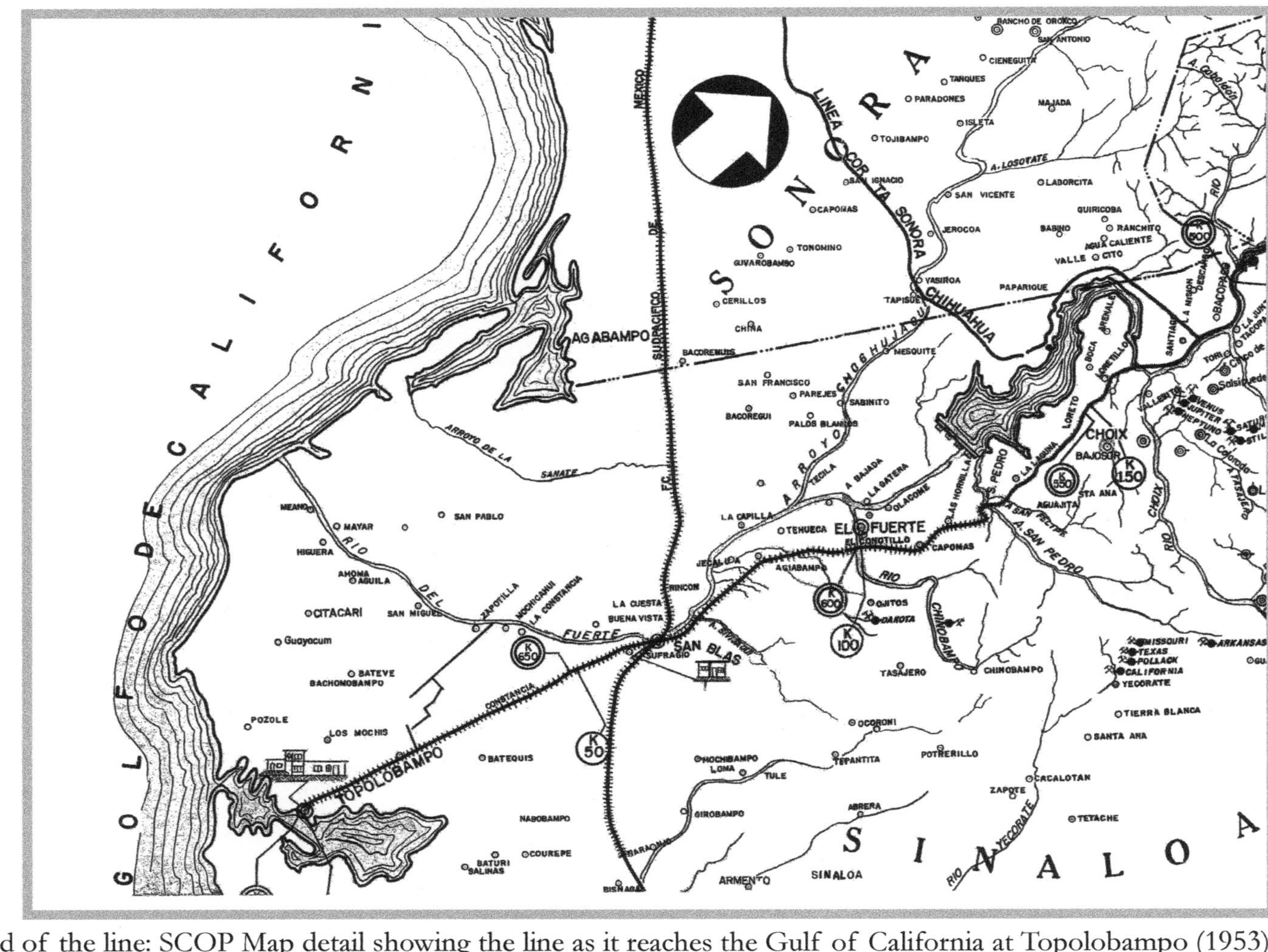

2.9. End of the line: SCOP Map detail showing the line as it reaches the Gulf of California at Topolobampo (1953).

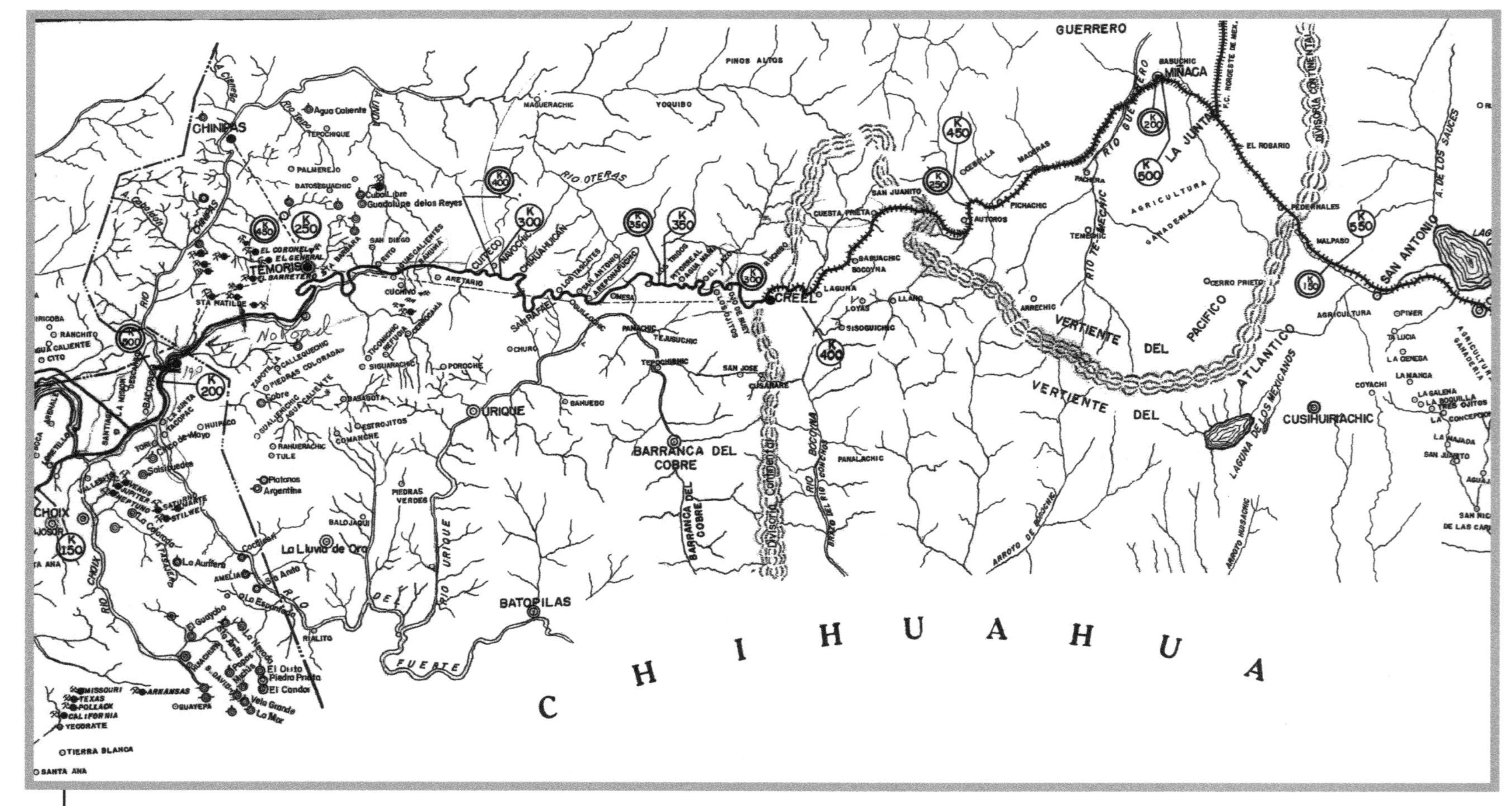

2.10. The Sierra Challenge: SCOP Map detail showing the line as it crosses the Sierra via La Junta, Creel, and Témoris (1953).

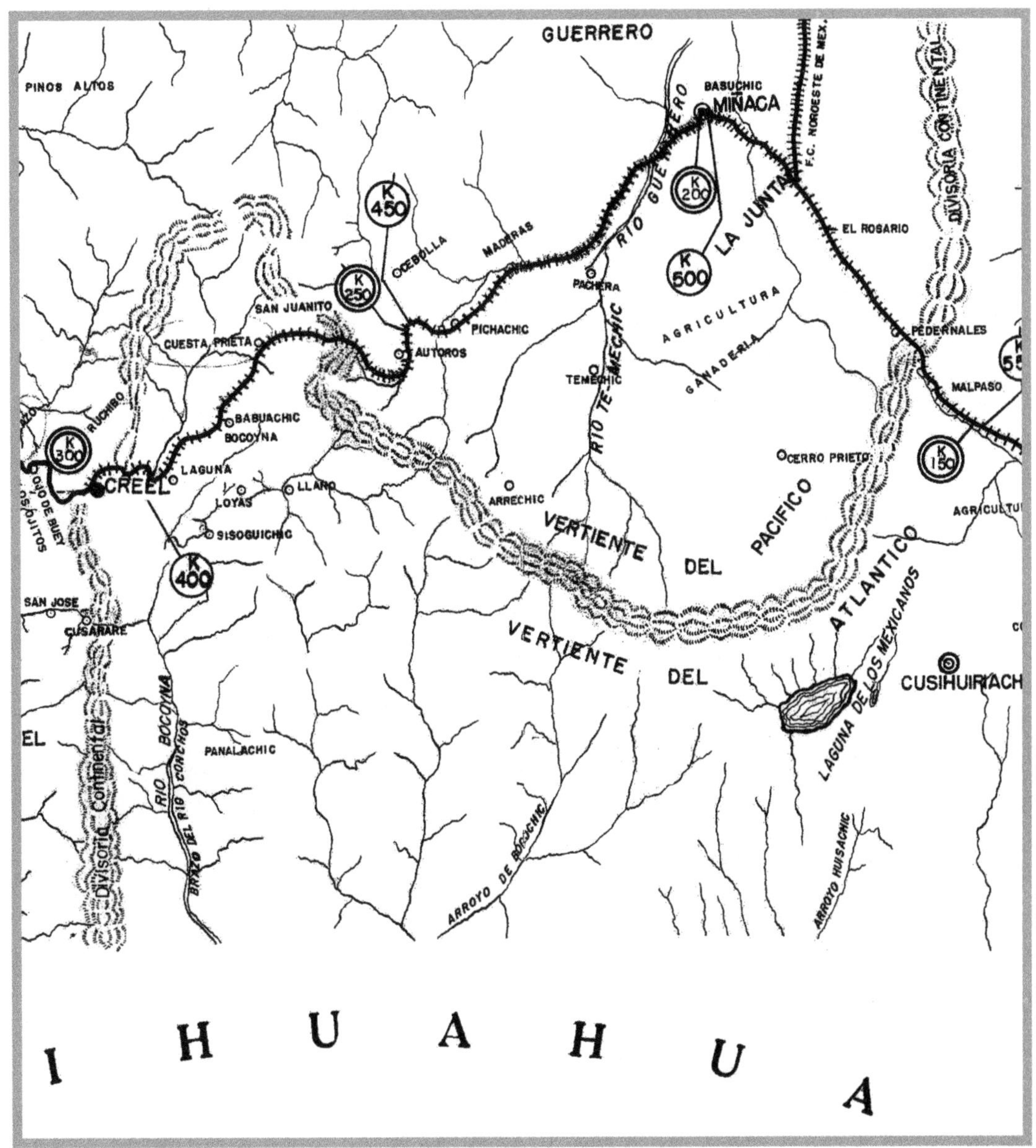

2.11. The challenge of the continental divide: SCOP Map detail showing the line from La Junta west to Creel (1953).

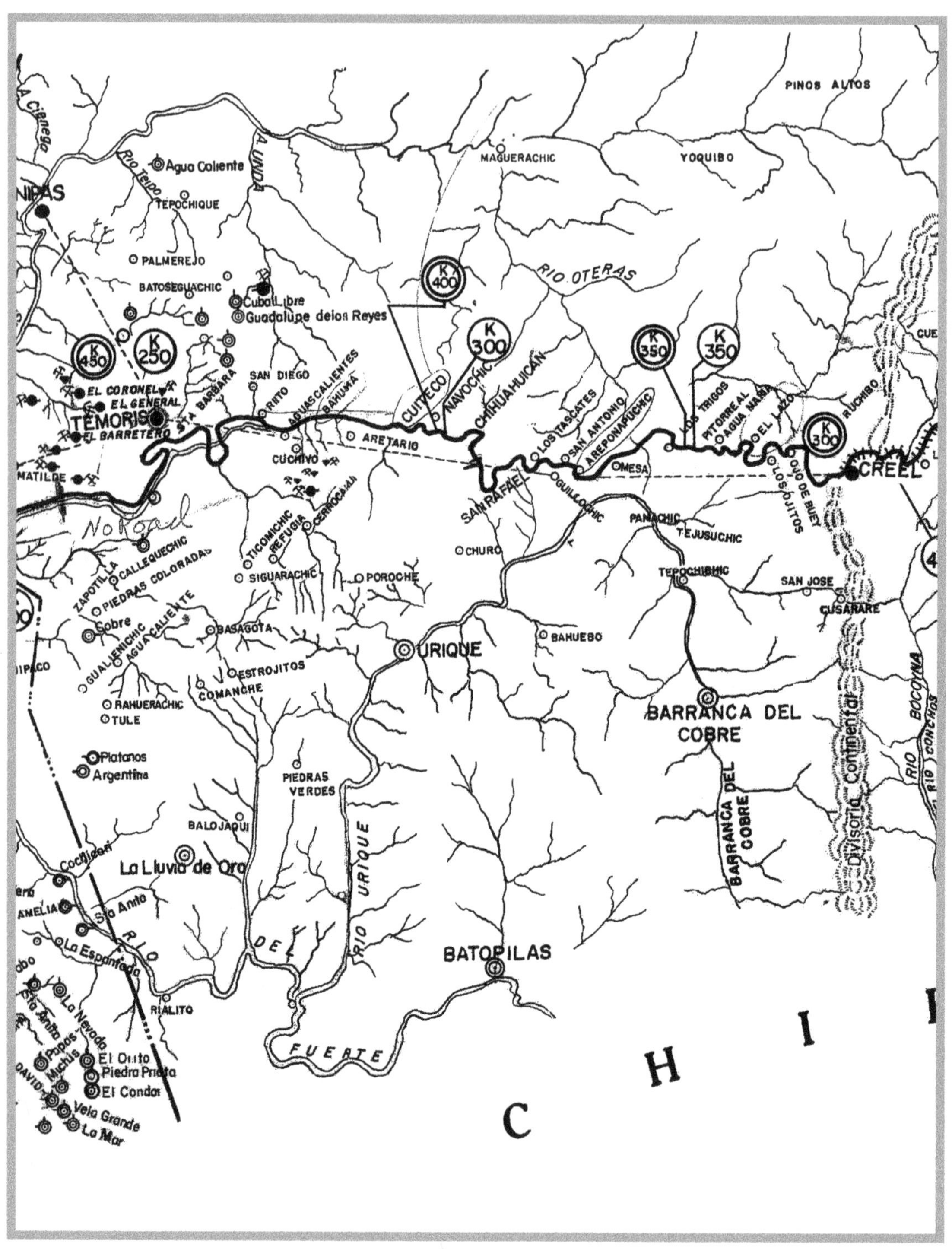

2.12. The challenge of the barrancas: SCOP Map detail showing the line from Creel west to Témoris (1953).

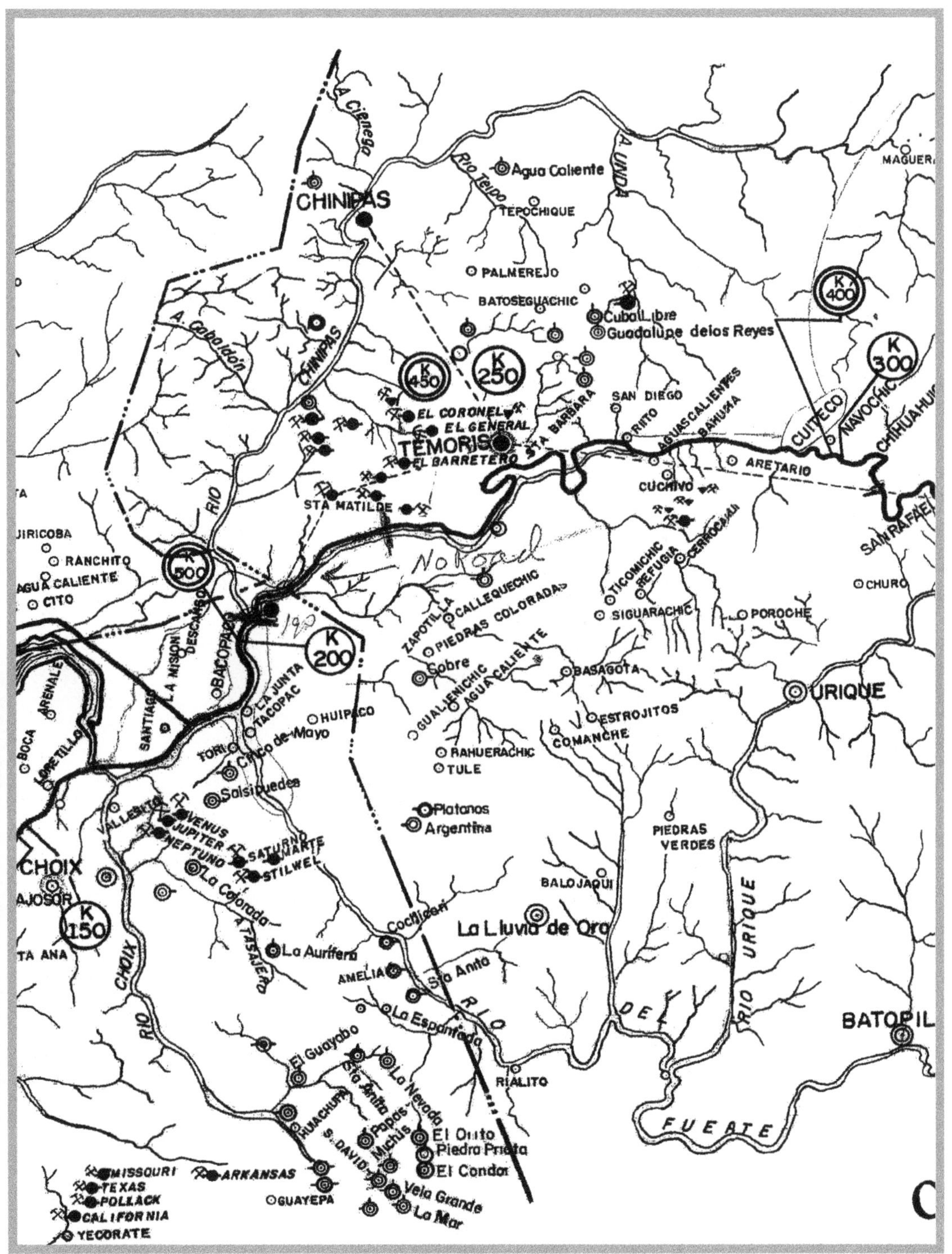

2.13. The challenge of the barrancas: SCOP Map detail showing the line from Cuiteco west toward Témoris and Choix (1953).

Before 1900

by Don Burgess

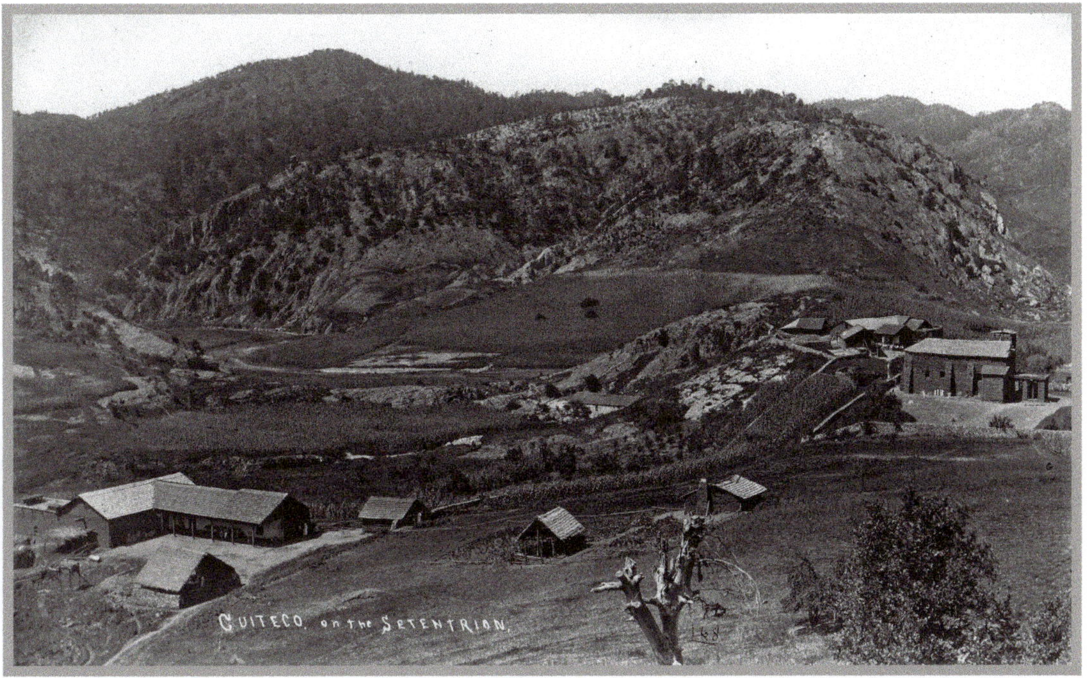

3.1. Cuiteco in 1887 (photo taken by Ira Kneeland on Owen's railroad survey trip). A plaque at the Balderrama's Hotel in Cuiteco quotes Owen as saying "Cuiteco is like a whirlwind of all the natural beauties that exist on the planet."

The church still stands but the entrance has been changed to the opposite end. The large house on the left is now a school. The present day train station is located on the other side of the creek in the mid left of the photo.

CHIHUAHUA

SALIDA AL MAR

TOPOLOBAMPO

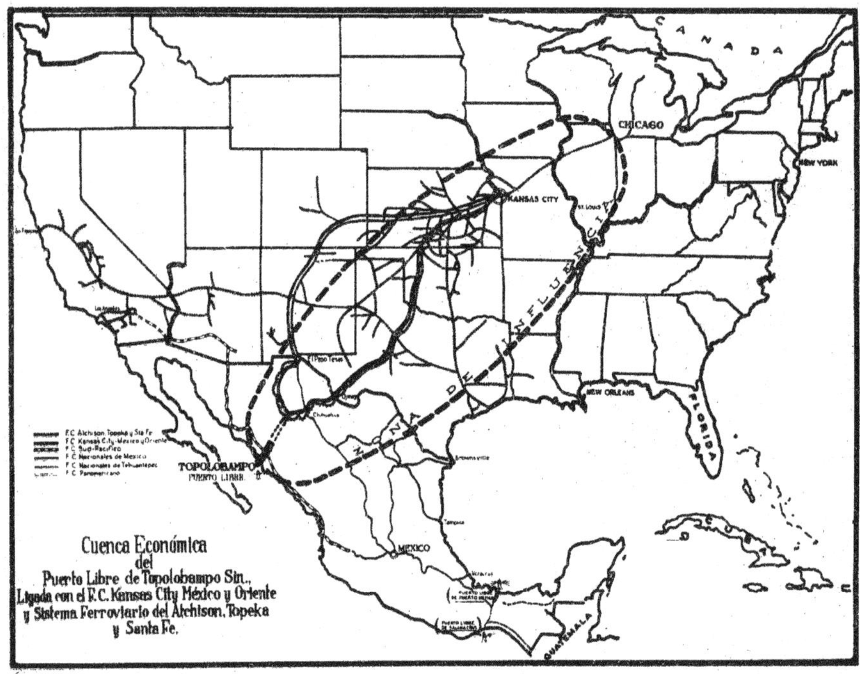

CONFERENCIA

S. M de G y E

ULISES IRIGOYEN

2 DE SEPTIEMBRE DE 1943

3.2. Cover of presentation given by Ulises Irigoyen in 1943. The map shows the planned links from Topolobampo to Chicago and highlights the railroad's zone of influence (courtesy of the Prof. Luís Urías Belderráin Library).

Since Dad's articles deal mostly with the construction period after 1900, I need to point out that the idea of connecting the eastern part of the United States by railway with Mexico's Pacific coast had been around for a number of years before 1900. According to the Chihuahua historian, Francisco R. Almada, the first person to come up with the idea of connecting Chihuahua with the Pacific Ocean by train was the Frenchman Hipólito Pasqueir de Doumartin in 1849. Along with the construction of a railroad to the coast, his idea was to colonize certain areas of Chihuahua with people from France (Almada, *El Ferrocarril de Chihuahua al Pacífico*, p.8.).

To put this in some historical perspective, note that this was just after the Mexican-American War (1846-1848), and a few years before the Civil War in the United States (1861-1865), and the War of the French Intervention in Mexico (1861-1867). And it was less than 50 years after the US bought the Louisiana Territory from France, and Lewis and Clark were sent by President Andrew Jackson to look for a water passage from the Mississippi River to the Pacific Ocean. The First Transcontinental Railroad was built in the United States between 1863 and 1869.

The last half of the 1800s was a time of railroad building in Mexico and a number of proposals were made for building railroads in the state of Chihuahua. The first train from Juárez arrived in Chihuahua City in December of 1882, and there were other lines reaching out into the mountainous mining and lumbering areas of Chihuahua.

One of the more interesting ideas, which never quite happened, was that of Col. James Reily, in 1862. I quote from Fred Coffey's unpublished Master's thesis, "Confederate Attempts to Control the Far West" (The University of Texas, Austin, 1930, as quoted in Dad's thesis.).

> As far back as the Civil War, Texans were thinking of a railroad to the Gulf of California. On January 26, 1862, Colonel James Reily, member of the Confederate Expeditionary Forces in New Mexico, after negotiating with Governor Luís Terrazas in Chihuahua City regarding a proposal for the State of Chihuahua to join the Confederacy, wrote:
>
> *"Chihuahua is a rich and glorious neighbor and would improve by being under the Confederate Flag. There are no such mines in the world as are within sight of Chihuahua City, but not developed for want of a stable government. We must have Chihuahua and Sonora -- by a railroad to Guaymas we render our great State of Texas the great highway of Nations."*

One of the persons to first consider a railroad from the USA through Chihuahua City and across the mountains to the Bay of Topolobampo, the route of the present day Chihuahua al Pacífico Railroad, was the civil engineer Albert

Kimsey Owen (1847-1915). Owen had been in Mexico on a railroad survey crew in 1872 where he became excited about the construction of a railroad across the Sierra Madre Occidental to Topolobampo, connecting the eastern part of the USA with Asia. The Bay of Topolobampo had already been recognized for its potential. It averaged 270 feet deep and could hold the entire United States fleet. And it was 400 miles closer to Kansas City than San Francisco.

In 1881, Owen incorporated the Texas, Topolobampo and Pacific Railway and Telegraph Company. Former US president Grant was among its officials. Owen saw railroads as a cure-all for the world's problems. He said: "When armies and legislation are powerless, the locomotive engine does not fail of success" (Kerr, p.36). A team was sent to Topolobampo to evaluate the harbor, the Fuerte River valley, how to get across the mountains to Parral via Norogachi, and to find the best route on to Eagle Pass, Texas. Included in the team were a U.S. businessman, the U.S. Consul, and an engineer. Their reports were glowing as to the possibilities, but no rails were ever laid (See *The Texas, Topolobampo & Pacific Railroad and Telegraph Company: Reports.*). Owen himself made a survey trip by mule across the Western Sierra Madres in 1887. He was impressed by the beauty of the mountains, the Tarahumara Indians, and the complete silence where one could feel the presence of God (Robertson, p.93).

Connected with his plans for a railroad, Owen became involved in socialistic-communal colonies of Americans in the Topolobampo area. In the first few years, over 600 Americans lived there, but by 1900 things had pretty well regressed to where they had been before. The harsh realities of pioneer life, disagreement among the settlers and "the unexpected obstacles confronting promoters in Mexico" (Pletcher, p.293), had taken their toll.[1] Owen's house in Los Mochis is now a cultural center.

The different people who developed plans to build a railroad across the Sierra Madre were not always realistic, but they were certainly dreamers and promoters. For example, in the survey report to Owen in 1881, Dr. B.R. Carmen said that "every mountaineer, every Indian, has mines. They are found on every hill. Build your road here, and you will have a town in every gorge" (p.32), and John Price added: "...when the grading for it (the railroad) is done many valuable properties will be unearthed" (p.40). Almada quotes a speech given by one of Owen's associates, Jesús Escobar, in the *Seminario Oficial*, the official communication of the State of Chihuahua (in 1875), where Escobar lays out the possibility that the railroad could be completed in one year (It took 86 years.). He also said that the people of Chihuahua could make use of the railroad to visit the Philadelphia Fair, and that Chinese and Japanese scholars could make use of the line to go from

1 Also see T.A. Robertson. *Utopia del Sudoeste (Una Colonia Americana en México)*. Also in English.

Topolobampo to Austin, Texas, to observe Venus passing in front of the sun (Almada, *El Ferrocarril Chihuahua al Pacífico*, p.51), which occurred in 1882.

People all along the route had great hopes of what the railroad would bring.[2] On April 2, 1913, the local newspaper at Alpine, Texas, *The Avalanche*, carried this article:

> **First Orient Train**
>
> While there was no blaring of trumpets, no blowing of whistles nor spilling of "shampoo" water, the orient railroad, the great northern outlet for this western country was completed to its depot in Alpine—its southern terminus for probably years to come—Tuesday evening.
>
> No silver spike was used to pin down the last rail, but notwithstanding, it like all the other parts which go to make up this magnificent system of transcontinental line is of the very best workmanship and engineering skill.
>
> Throwing open to settlement, as one might say—for West Texas is yet in her infancy—a virgin country full of possibilities for the capitalist as well as the settler, there is no reason not to believe that the population of this section will soon rival that of our sister counties in the central and other portions of Texas.
>
> Alpine, the southern terminus of the orient, offers opportunities to the capitalist that will not long remain unaccepted. Coupling this with the fact that God has endowed us with a climate that is not excelled anywhere and seldom equaled, there is no reason why this should not be the Denver of Texas. (Casey. *Alpine, Texas then and now*, p.250).

The Director of Railroads of Mexico, Ulisis Irigoyen[3], had similar ideas for Chihuahua City. He had gone on a railroad survey trip across the Sierra in 1939 and in a report presented to the Sociedád Mexicana de Geografía y Estadística on September 2, 1943, entitled "Topolobampo, Salida al Mar,"[4] he said:

> It is not a utopian idea to think that once the Kansas is finished, the population of the State of Chihuahua will increase ten times, the same as the value of property and multitudes of industries will be established. The capital of the state will become the largest border city and of the greatest commercial and industrial importance in the North of the country.

2 For a list of the major promoters of the railroad, beginning in 1850, and also of the engineers who worked on the final construction, see Francisco Almada, *El Ferrocarril de Chihuahua al Pacífico*, pp.174-178.

3 The village of Irigoyen on the rail between Bahuichivo and Témoris was named after him.

4 My thanks to Luís Urías for providing a copy of this privately printed issue.

Map Drawn in 1887 from the Owen Survey

Note that the railroad was to go through the mining town of Cusihuiriachi. The actual line goes about 30 miles to the north through Cuauhtémoc (A branch line was made from Cuauhtémoc to Cusihuiriachi in 1911.) Also note the two possible routes shown in the area of Bocoyna and that two Bocoynas are mentioned. The one to the right would be the present day Bocoyna on the Concho River headwaters. The map is somewhat vague as to exactly where the line goes after Cuiteco to get out of the mountains, but Robertson quotes Owen as saying that the line would go through the Septentrión Canyon, which is where the final line was built.

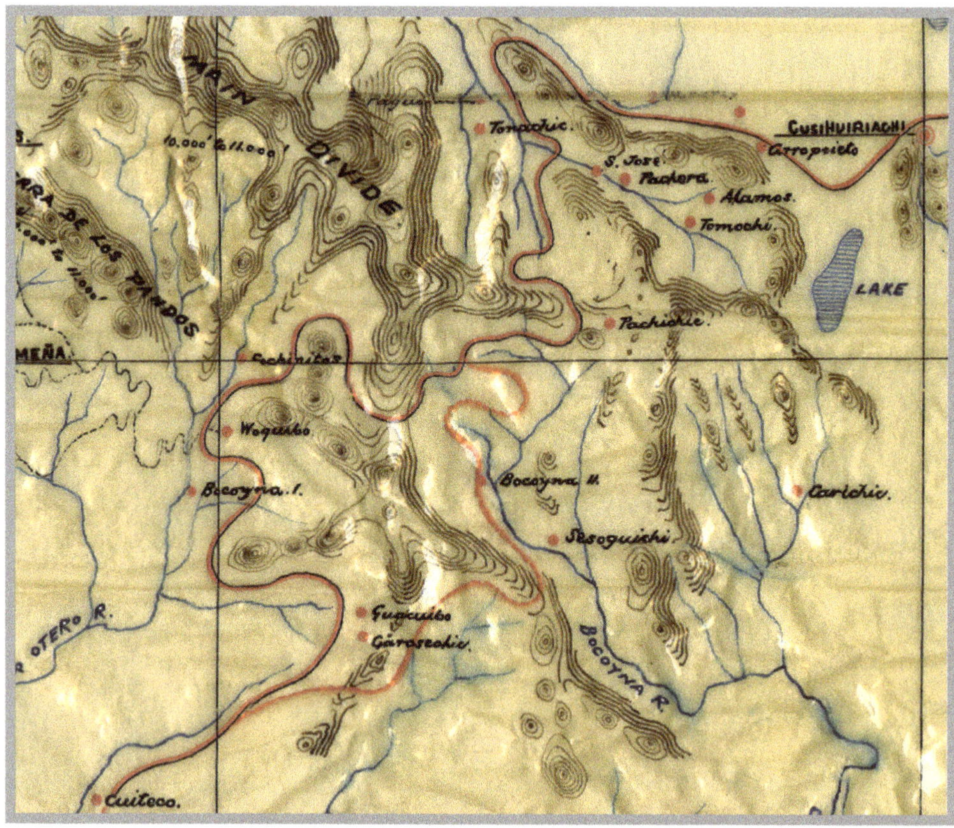

3.3. Map from the Owen Survey by Alfred Rosenzweig, 1887: Section showing two possible routes across the Main Divide in the area of present day San Juanito and Creel.

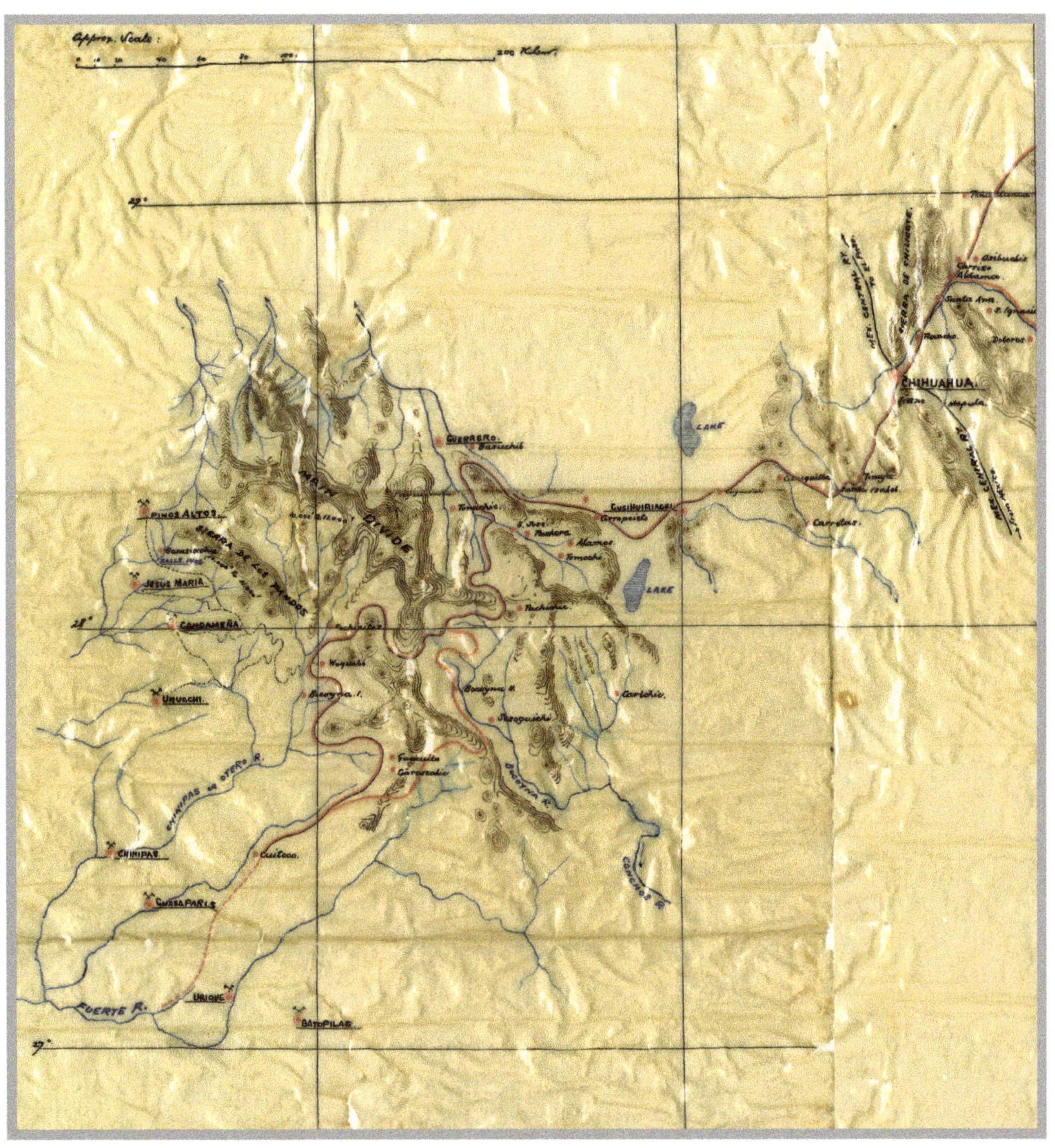

3.4. Map from the Owen Survey, 1887: Detail showing the proposed routes through the canyons of the Sierra in Chihuahua and Sinaloa.

A quick look at the description of the Urique Canyon written by the Jesuit priest Juan Salvatierra in 1684 gives an idea of the problems to be faced in getting a railroad across the Sierra Madre Occidental, problems that some of the promoters did not understand:

> On discovering the precipices, such was my terror that I immediately asked the (Tarahumara) Governor if it was time to dismount, and without waiting a reply, I did not dismount, but let myself fall off on the side opposite the precipice, sweating and trembling all over with fright; since there opened, on the left, a chasm whose bottom could not be seen and, on the right, some perpendicular walls of solid rock. In front was a drop of four leagues, at least, not a gradual slope but violent and steep, and the trail, so narrow that at times it is necessary to travel by jumps, for not having a place in between to put your feet. (As quoted in Francis Javier Alegre, S.J., pp.67-68.)

Even after making a survey trip across the mountains on horseback, Owen was extremely optimistic about what it would take to complete a line. He said: "We encounter few problems for the construction of the line through the Septentrión Canyon, from La Guaza to Bocoyna. Tunnels can be avoided. The grades will be moderate, it all promises a lower cost than the construction of the lines that cross the Sierra Madre in the United States" (Robertson, p.94). His optimism can be seen in the expectation that tunnels could be avoided. In fact, 88 had to be built.

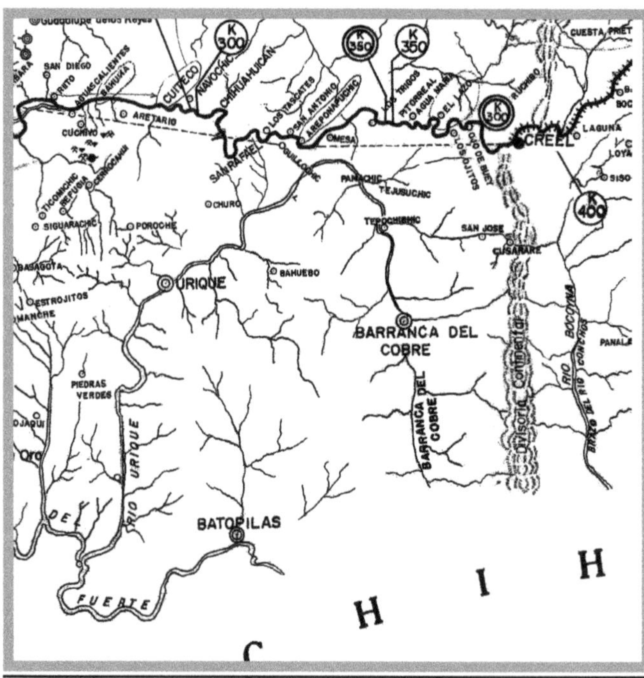

3.5. SCOP Map: Detail showing many rivers cutting canyons through the sierra. Note the Urique river that cuts the Barranca del Cobre through the mountains to the west of the continental divide before joining the Fuerte River. To the north of the line are the beginnings of the Oteros-Chínipas River Canyon. (1953).

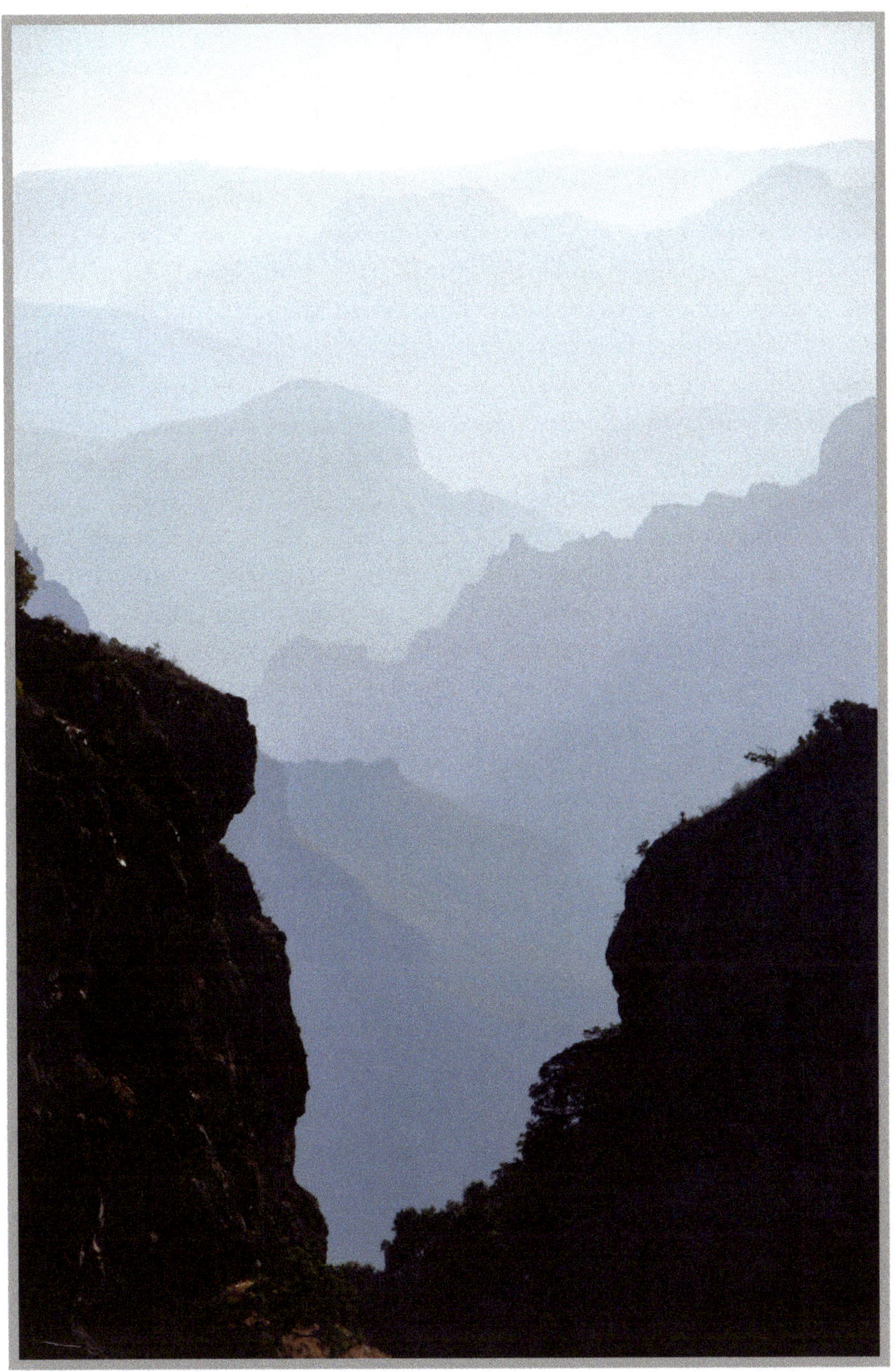

3.6. Urique Canyon near where Salvatierra saw the canyon in 1684 (photo D.B. c.2010).

And with that introduction, we will go on to my dad's articles covering the work of the "Prince of Promoters" Arthur Edward Stilwell in the early 1900s and the final completion of the railroad by the Mexican government in 1961. To set the stage, I will quote from Stilwell's book, *Cannibals of Finance, Fifteen Year's Contest with the Money Trust* (pp.126-127). He wrote:

> Before I finish my chapter, I want to picture to my readers the Orient road as I see it: The Kansas City, Mexico and Orient Railroad is one of the greatest enterprises of today; it opens an empire of wealth; it opens one of the treasure houses of the world in the mines of Mexico. It will build a port that will rival any on the Pacific coast; it shortens the line across the continent; it makes a great short-cut to the west coast of Mexico, Central and South America. On the line of this road, when finished, there will be three smelting centers—the smelters now at Chihuahua, smelters at the border and a smelter at the coast. The lumber of the Sierra Madre Mountains will find a market as far north as northern Oklahoma. It will supply all the ties for western Texas. Along the line of the finished road will be great plants treating the ore from the dump heaps placed there in the years of long ago; along the road will be two or three cities which will equal Cripple Creek as mining centers.
>
> The Port of Topolobampo will be one of the great cities of the Pacific; it will have its line of steamers to the Orient, Central and South America, New Zealand and Australia. The early vegetables, oranges, etc., which are one month earlier than those in California, will come in train loads to Chicago and eastern markets, and the one hundred miles of the Fuerte River Valley, as rich as the valley of the Nile, will contribute great earnings to the road. The hundreds of thousands of acres of level land east of Chihuahua will be irrigated, and the cultivation of sugar beets and cotton furnish northern Mexico with all these products it can use.[5]

Many things have happened since Dad's articles were published. In 1996, for example, the Chihuahua al Pacífico Railroad was sold by the Government of Mexico, for 255 million pesos, to a group made up of Grupo México, S.A. (74%), Constructura ICA (Ingenieros Civiles Asociados) (13%), and the Union Pacific Railroad Company (13%) (*El Heraldo de Chihuahua*, June 27, 1997). But a discussion of the operation of the railroad in recent years will have to be the topic of another book.

5 It is interesting that Stillwell did not mention tourism. For photos of Stillwell, Owen, Creel, etc., see Kerr.

Glenn Burgess as a Promoter

Glenn, himself, I should point out, was a promoter, as can easily be seen in his articles.

After graduating from the University of Texas at Austin with a degree in business administration and journalism, he worked with the Conoco Travel Bureau in Denver from 1931-35, in 1946 as manager of the first radio station in Alpine, Texas, and for sixteen years in the field of Chamber of Commerce Management in different places, including Alpine, Texas, Santa Fe, New Mexico, and St. Joseph, Missouri. During World War II, he was a Lt. Col. in the Texas State Guard, helping to guard the Texas-Mexico border in the Big Bend area.

At some early point in his life, he became interested in photography and, as a photo-journalist, he was one of the early promoters of the Big Bend National Park in West Texas. His photographs of the Big Bend can be found in the Archives of the Big Bend, Bryan Wildenthal Memorial Library, at Sul Ross State University in Alpine, TX.

The Construction

news articles by Glenn Burgess with notes by Don Burgess

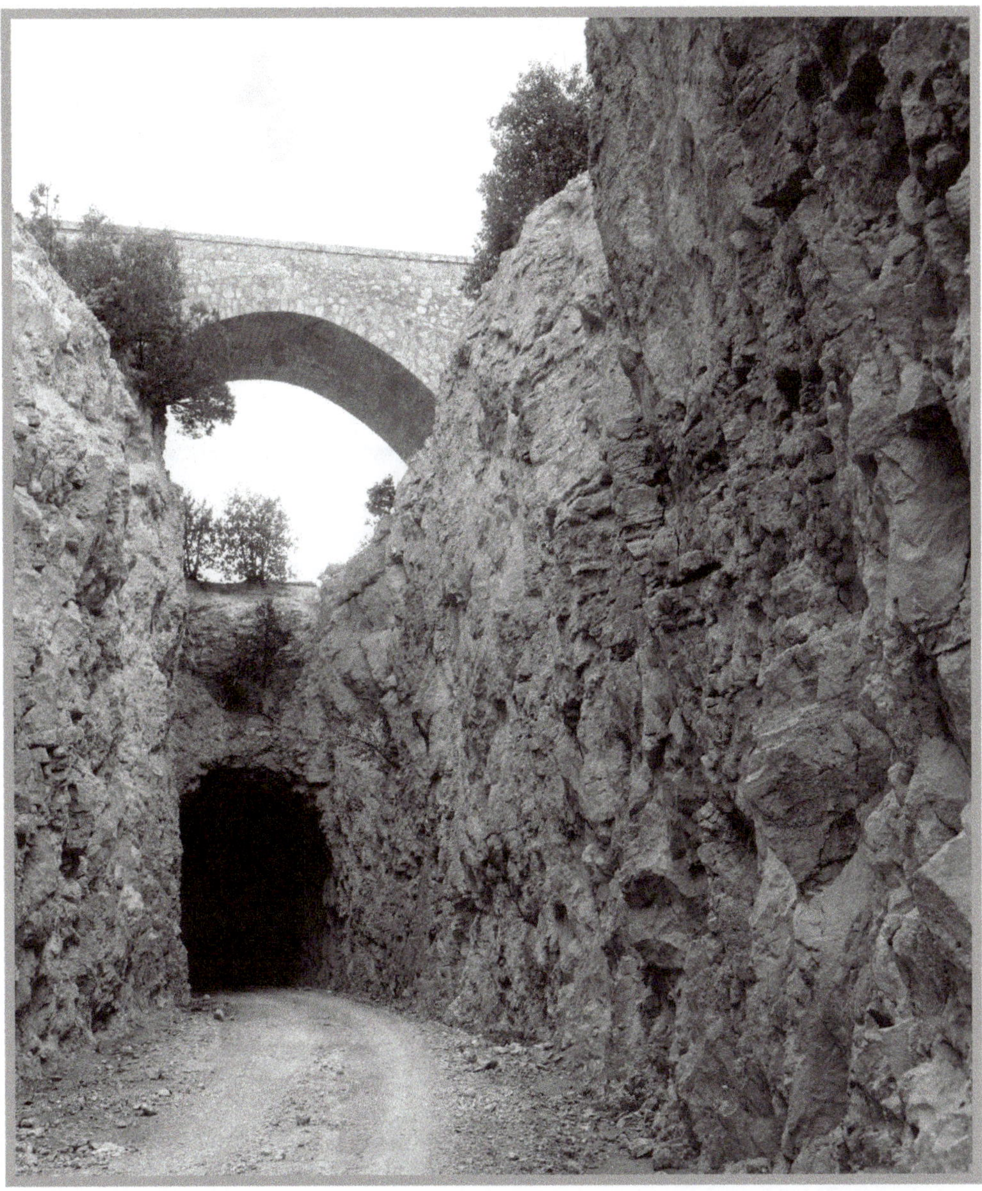

4.1. El Lazo, where the railroad loops over itself, between Creel and Pitorreal, was the idea of Ing. Francisco Togno (photo G.B. 1955).

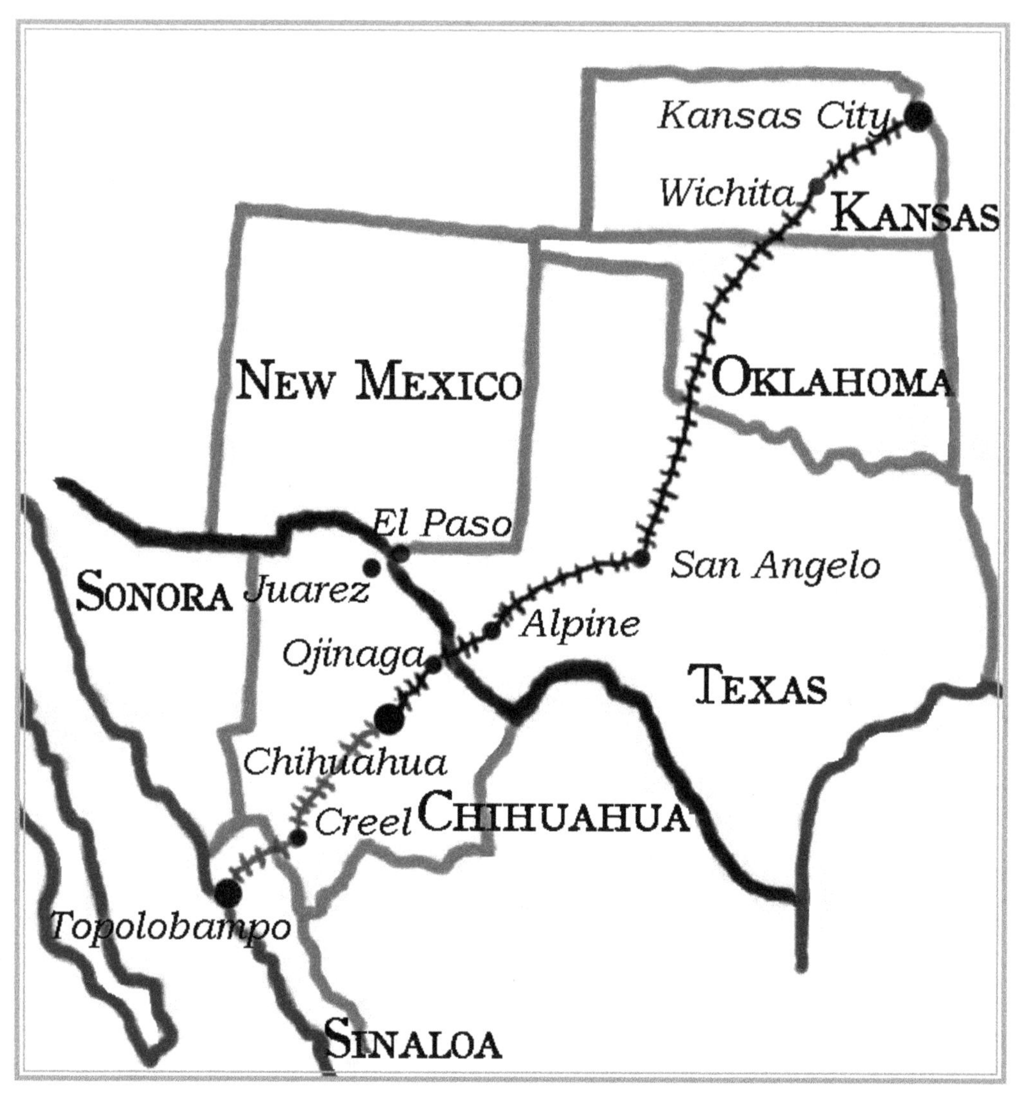

4.2. Sketch of the planned Kansas City, Mexico & Orient Railroad.

The El Paso Times

- Sunday, May 19, 1957

Chihuahua-Pacífico Railway To Open Mexican Resources

by Glenn Burgess
Times Correspondent

During the past three years, news has been circulating in the El Paso area that Mexico was working on the completion of a railroad from Chihuahua to the Pacific. This line could be considered an extension of the old Kansas City, Mexico & Orient and the Noroeste de Mexico that connect with Juárez via Casas Grandes.

Last October President Ruíz Cortines, on his visit to Chihuahua City, announced the railroad would be completed soon.

The new line will open up vast forest, mine, agricultural and scenic resources—country that only a few hardy people from the United States have seen.

What President Ruíz Cortines was referring to is a gap through the Sierra Madre of 176 twisting and rugged miles which today has no rails on it.

This is one of the three major railroad construction projects in Mexico, and one of few in North America.

The project started in 1900 when the idea of a shorter route to the Pacific from Kansas City was conceived by A.E. Stilwell.

Stilwell had just promoted the Kansas City Southern and lost it to John W. (Bet-a-Million) Gates.[1] To get Stilwell out of a period of mental depression, friends held a morale building banquet for him.

SHORTEST ROUTE

The banquet caught Stilwell on the rebound. Apparently without any forethought he calmly announced he had a bigger and better project. Using a piece of string, he demonstrated on a map of North America that the shortest route from mid-continent was from Kansas City to the Gulf of California at Topolobampo.

The new line was to be 1,659 miles in length and 400 miles shorter than any other from Kansas City to the Pacific. The Mexico section would be 598 miles, connecting with Texas at Presidio.

1 This line ends at Port Arthur, Texas, which was named after Arthur Stilwell.

Pancho Villa

Friedrick Katz quotes Stilwell as saying: "Pancho Villa…had been one of my contractors. He had twelve teams. I used to let him a mile or two of work and when he had finished I would let him another mile or two" (p.69). But things were apparently not too amiable between them. Richard Fowler states: "While the Orient was being built, he (Stilwell) snubbed one of his smaller contractors (Pancho Villa) by never inviting him into his private car" (p.106).

The railroad certainly played an important role for Villa. His wife Luz Corral was from the railroad town of Riva Palacio (San Andrés) (see Katz, p.148). And in 1916, shortly before Villa's raid on Columbus, New Mexico, his men took 19 Americans connected with the mine at Cusihuiriachi off the train and shot them. Only one escaped (See Katz pp.558-560; *Gringos' Curve* by Christopher Lance Habermayer.).

Speaking of Pancho Villa, when I was riding the train to Chihuahua City in 1965, I struck up a conversation with an older man who said he was a tailor and that during the revolution he had once made a suit for Pancho Villa. He said that, while working on it, he would go every day to Villa's house for measurements. But one day he did not go because of bullets flying by. When he went the next day, Villa asked him why he had not come. "Because of the bullets flying by," said the tailor. "You don't have to worry about the bullets that fly by," said Villa. "It's the ones that hit you that you have to worry about." The federal troops moved in the next day and Villa never wore the suit.

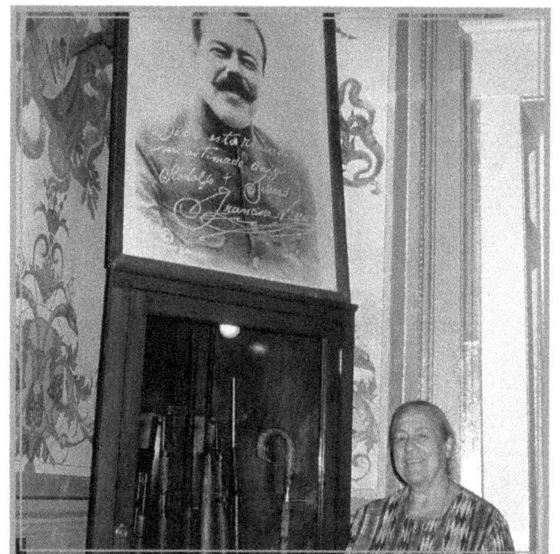

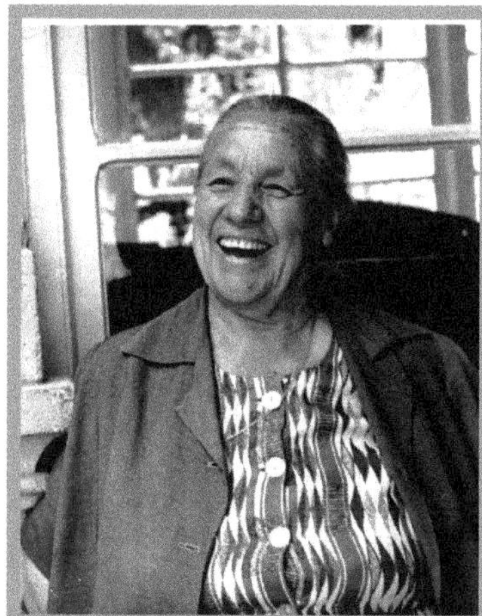

4.3 and **4.4.** Pancho Villa's wife, Luz Corral, in her Chihuahua City home (photos G.B. 1960s).

Two months after the banquet, Stilwell organized and by July 4, 1900, the first rail was laid at Emporia, Kans. By 1908, 729 miles was completed with an additional 241 graded. Pancho Villa was one of the grading contractors in Chihuahua.

Much of the financing came from England. The stock was watered to the point the outstanding securities doubled actual investment.

By 1912 Stilwell's engineers had not figured out a practical way to cross the mighty Sierra Madre and he lost control. William T. Kemper and others purchased the line about 1917 for $3 million. Government upheavals in Mexico and other troubles kept the project from being completed.[2]

With the discovery of oil in the McCamey-San Angelo area in 1923, the road showed signs of making a little money. The boom station of McCamey is said to have grossed $750,000 worth of business in one month. However, the rolling stock was worn out, the poor cross-ties, grades and bridges were causing financial trouble, and the Santa Fe purchased the line at a figure in excess of $14 million.

President William B. Story Jr., didn't figure the Mexican section very valuable and in 1929 sold 289 miles of rail and unconstructed portion of 215 miles to B. F. Johnston, a land and sugar operator, for $650,000 in cash and carried a mortgage of $900,000. Johnston defaulted and in 1940, the Mexican government took over from Johnston in conformance with its program of nationalizing railroads. Later, a token payment of $90,000 was made by Mexico in full discharge of the obligation.

TRACKS LAID

When the Santa Fe acquired the orient property, tracks were laid on the route from Kansas City to Alpine, Texas, via Wichita, western Oklahoma, Benjamin, Sweetwater, San Angelo and Fort Stockton, Texas. In 1930, the new company acquired trackage rights over the Southern Pacific from Alpine through Paisano Pass, then built a new line to the Rio Grande at Presidio. A special train went to Chihuahua City on Nov. 2, 1930.

After the depression, and with World War II pointing out

2 Irigoyen, in his report in 1943, blamed the Southern Pacific Railway and others in the United States for the KCM&O not being finished. He said they feared that the KCM&O, having a much shorter route to the Pacific, would take away their freight business. Irigoyen argued that, on the contrary, the effect would be just the opposite.

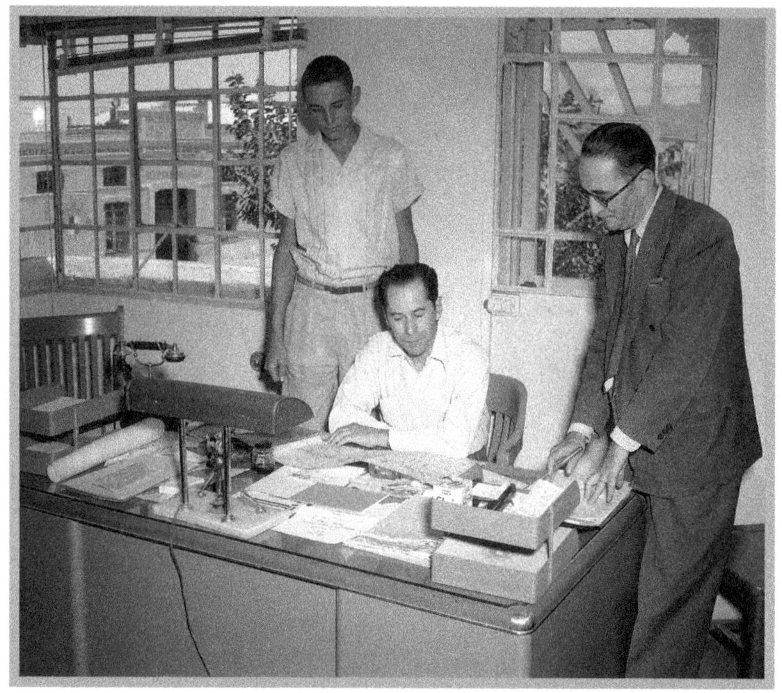

4.5. Alfonso Rincón Benítez (seated), head of construction and rehabilitation, talking with Don Burgess and Jorge Togno in the Chihuahua City office (photo G.B. 1955).

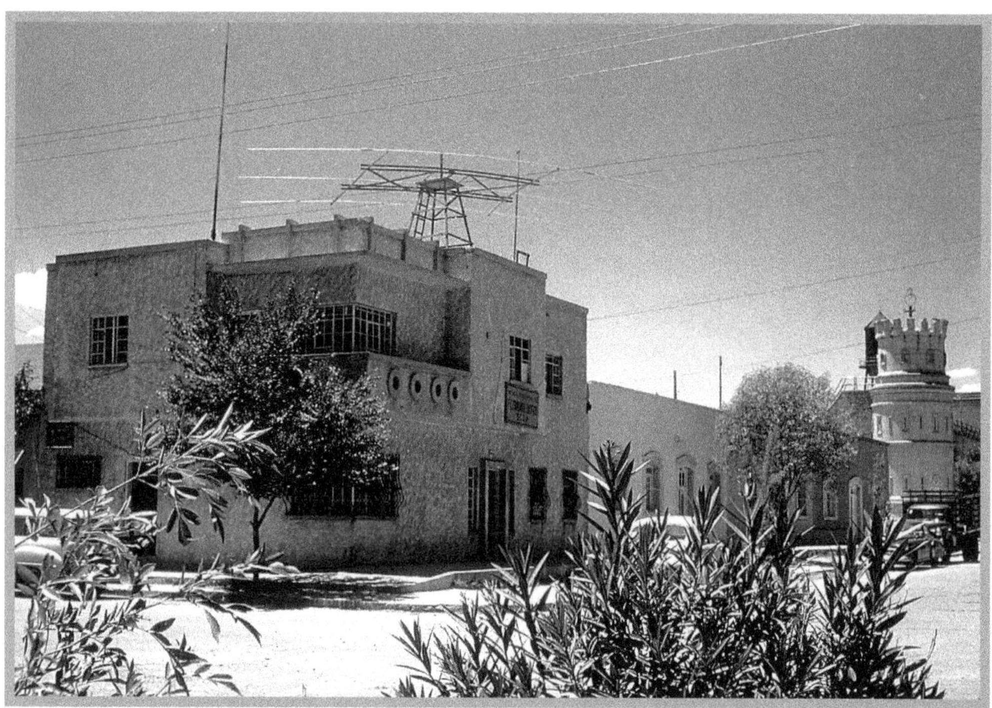

4.6. The Chihuahua City headquarters of railroad construction. The Chihuahua State Prison is in the background (photo G.B. 1955).

that Presidio was nearer to the Pacific than it was to Austin or Fort Worth, new interest in the line was expressed by the Mexican government.

Again, from Douglas, Ariz., to Guadalajara, Mexico, an airline distance of approximately 850 miles, Mexico had neither rail nor highway transportation across the Sierra Madre Occidental. Chihuahua and Hermosillo, and Durango and Mazatlán were short distances from each other by air but ground transportation met an impenetrable mountainous barrier from 6,000 to 10,000 feet in height and more than 100 miles wide. The future economical development of the nation demanded crossings.

Mexican civil engineers, led by Francisco Togno, and assisted by his brothers, Ramón and Jorge, ran line after line and finally decided completion of the old Kansas City, Mexico & Orient was possible. They also surveyed a route from Durango to Mazatlán.

When the Togno brothers went to work, they had 176 miles to complete between the western terminal of Creel in the western part of the state of Chihuahua, and eastern terminal of San Pedro in northeastern Sinaloa. Trains were running from Ojinaga to Chihuahua City to Creel, a distance of 352 miles, and from Topolobampo to San Pedro, a short distance of 68 miles.

Only 176 miles was left for completion, and gradually, this stupendous engineering and construction feat is closing the gap between Creel and San Pedro. Many think President Ruíz Cortines was overly optimistic about completion dates. But the need is too great, and the amount of money already invested in the construction of grades, cuts, fills, tunnels and bridges is too much to stop now.

The dream of the not too practical promoter, A.E. Stilwell, is drawing to a rapid fulfillment. Many of the benefits he predicted will also become a reality.

GETS NAME

The new line will be known as the Chihuahua al Pacífico, and for some time it has been operating under that name. It is a government owned and operated project. The construction work is under the direction of the Secretary of Communications and Public Works.

Francisco Togno, civil engineer, is now in charge of all railroad construction work in Mexico. The Chihuahua al Pacífico is undoubtedly his pet project. He knows every foot of it.

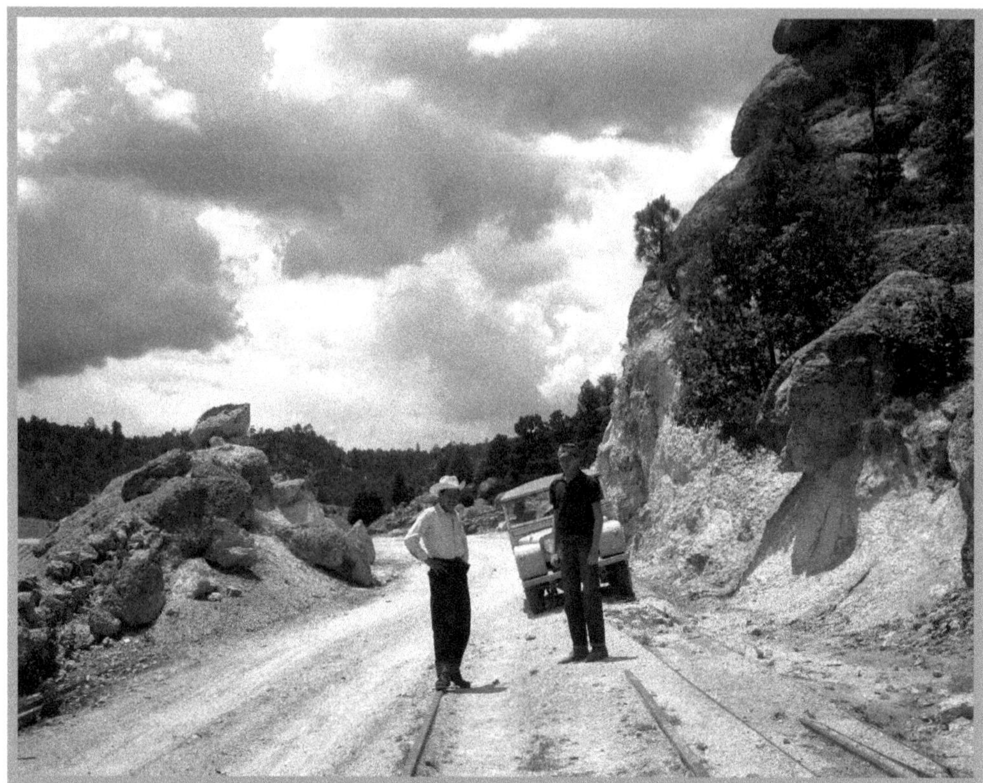

4.7. **INTO SIERRAS**—From the end of the line, the Chihuahua al Pacífico plunges into the rugged volcanic Sierra Madre Occidental. The road bed makes a good highway at present. Scene is west from western side of Creel (photo G.B. 1955).

Creel

During the early 1900s, the name "Chihuahua al Pacífico" was already being used for the line going west from Chihuahua City that was being promoted by the Governor of Chihuahua and Mexico's Ambassador to the USA, Enrique Creel.

The father of Enrique Creel, Reuben, was a native of Greensburg, Kentucky. During the Mexican American War (1846-1848), he served as an interpreter. After the war, he remained in Chihuahua, marrying into the Cuilty family (of Irish descent). He was appointed as US Consul to Chihuahua by President Lincoln (1863-1866).

Governor Creel, after whom the town of Creel was named, was married to a daughter of Luís Terrazas, one of the world's largest land owners. Creel became the vice-president of the Kansas City, Mexico & Orient.

Before the founding of Creel, the Tarahumara ranch located there was called Segórachi "Place of the Tadpoles". Another ranch nearby was called Rochibo, which refers to a flat area where an edible wild green grows known as rochíware.

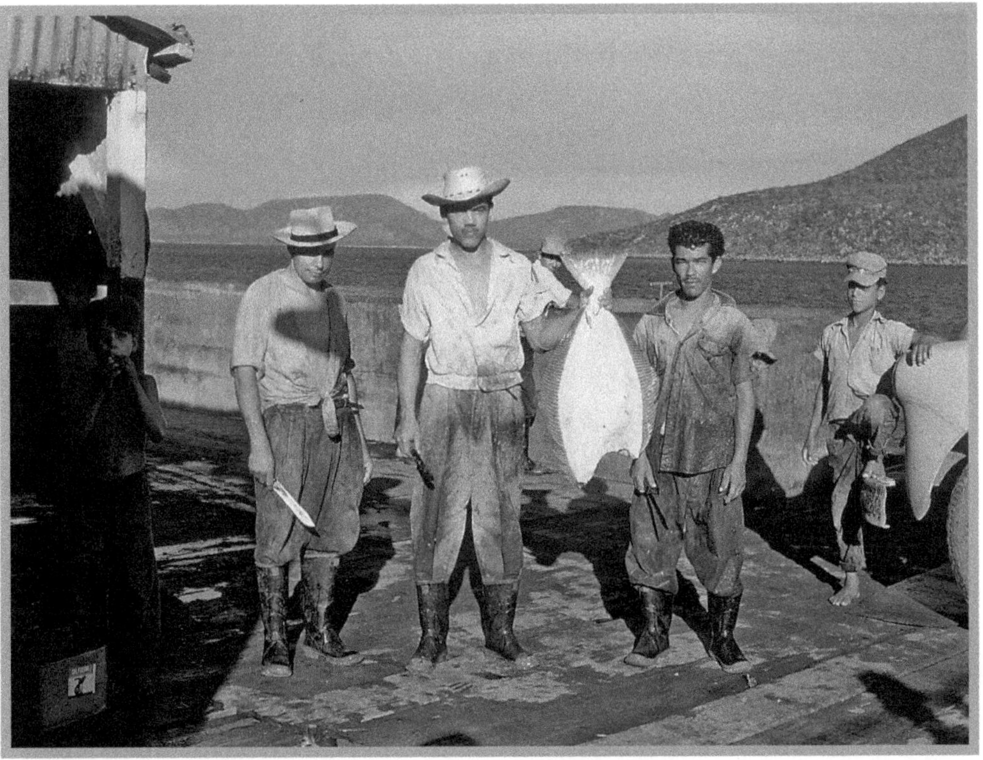

4.8. **END OF LINE**—The end of the line is the Bay of Topolobampo—largest natural harbor between San Francisco and Chile.

FRESH FISH—Fresh fish from the Pacific will be shipped over the completed railroad to El Paso and Mid-Continent cities. Displaying a halibut are two workers from a freezing plant at Topolobampo—the 11-mile bay in the background (photo G.B. 1955).

How the Canyons of the Sierra Tarahumara were Made

Geographer Robert Schmidt describes the country in the following way:

> The Sierra Tarahumara is part of a large uplifted volcanic plateau. As a result, the Sierra does not exhibit a classic crest or watershed divide. The upper portion of the Sierra consists of great thicknesses of extruded volcanic rock, typically 1000 to 1500 feet thick. Much of the upper volcanic material is ash flow tuff which was formed during the mid-Tertiary period. These voluminous quantities of volcanic rocks are laid down over folded, early Cretaceous and older sedimentary rocks. The roughness of the terrain is the result of headward erosion, and down cutting by youthful, deeply incised rivers. As a result [and as a result of plate tectonics], the Sierra Tarahumara probably possesses the world's largest proliferation of great canyons confined to a relatively small area. Because the Sierra Madre Occidental is a major barrier to moisture laden winds from the tropical eastern Pacific Ocean, the mountains are an extremely important source of water for the very productive irrigation districts on the surrounding lowlands. (From Schmidt's map of the Sierra Tarahumara. International Map Co. UTEP, El Paso, Texas, an earlier edition.)

The Tarahumara have an interesting way to describe how the canyons were made:

> They say that long ago Golachi, the crow, made the canyons. The land used to be flat. There were no canyons to carry off the water. That is why they say Golachi walked across the land. In this way, places were made so that the water could run off. That is why there are many canyons today. In parts they are deep and ugly.

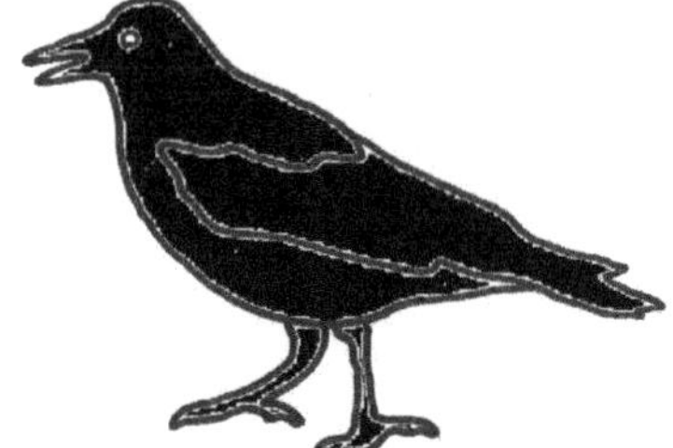

FORT WORTH STAR-TELEGRAM

- August 20, 1957

Effect Seen on Texas Economy

Work on Two Mountain Railways In Mexico Fast Nearing Completion

by Glenn Burgess

The people of the United States have forgotten their railroad building days, but not so in Mexico.

Steady, persistent work on at least two mountain scenic railways is bringing to completion transportation dreams that began in the days of President Porfirio Díaz.

Completion of these railroads will affect the economy of Texas, especially West Texas, and mid-continental United States.

The old Kansas City, Mexico & Orient, now known in Mexico as the Chihuahua al Pacífico, or Chihuahua to Pacific, will link West Texas with the Gulf of California. To accomplish this, 175 miles of construction is required, involving a series of cuts, fills and tunnels through rugged mountains.

Faraway Gulf of California isn't so far when the map is studied. Topolobampo, the western terminus of the railroad, is as close to Presidio, Texas as is Fort Worth. Mazatlán, the gem of the Pacific, below the Tropic of Cancer, another link in the system, is as close to Alpine, as is Galveston. The Barranca del Cobre, an abysmal canyon that is at least 6,000 feet deep, is nearer Presidio than is San Angelo.

TO SERVE VAST AREA

This new railroad will serve the largest remaining virgin stand of ponderosa pine and oak in North America; fabulous mines of copper, silver and other metals; one of the continent's richest agricultural empires; two of the finest ports on the Pacific; tourist meccas, both in the mountains and on the Pacific, and a fishing industry that includes the largest freezing plant in the world for shrimp.

Cutting the distance from Kansas City to the Pacific by 400 miles, it is possible this

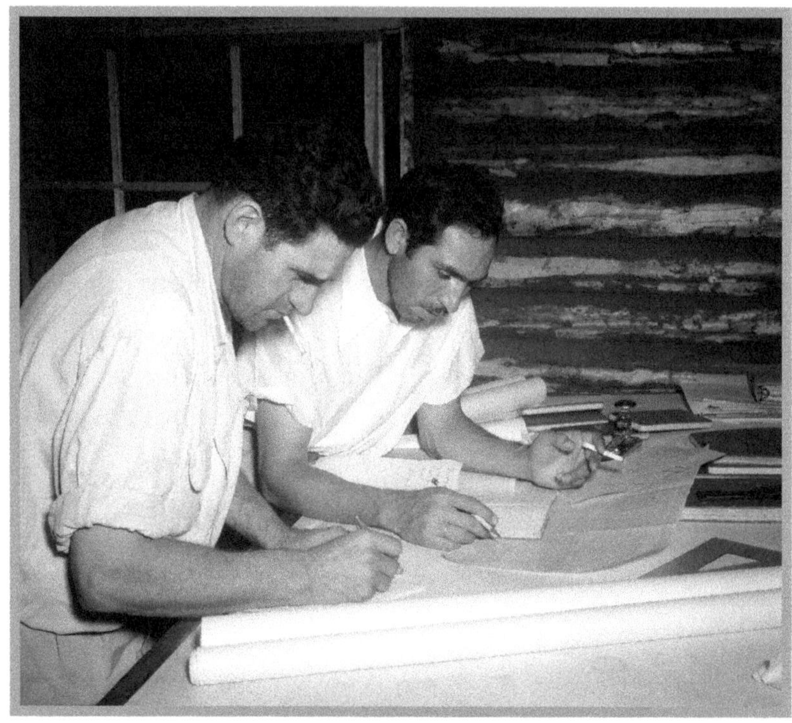

4.9. Ing. Duarte and unknown engineer (perhaps David Ávila) working on plans (photo G.B. 1955).

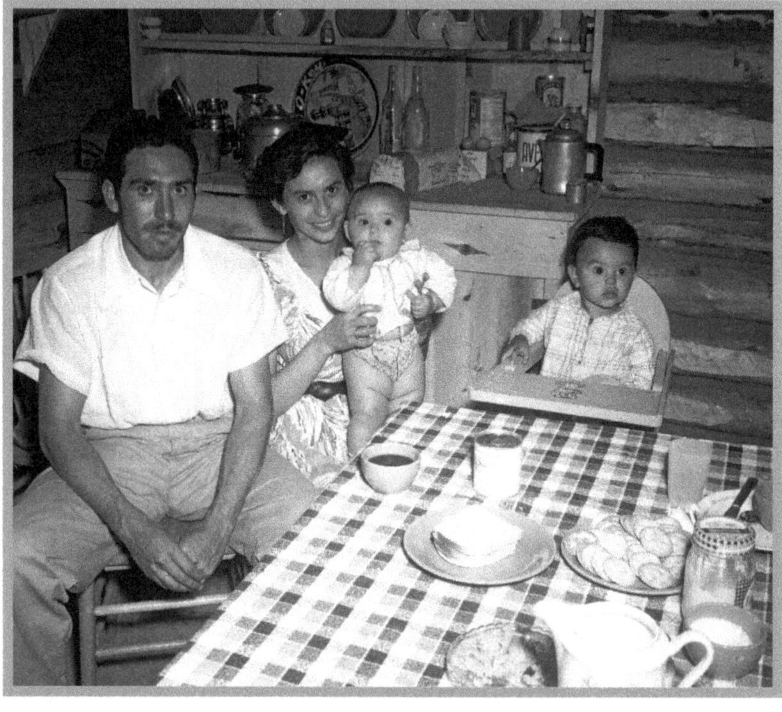

4.10. The families of the engineers often lived in the camps. Here are seen José Duarte Espinoza and his family, probably at Areponápuchi (photo G.B. 1955).

railroad will start a stream of freight from the Orient. It will carry fresh vegetables, fruit and fish to the produce markets of St. Louis and Chicago.

Years ago, Arthur Stilwell, a railroad promoter, got the map out and discovered he could reach the Pacific by building a railroad southwest from Kansas City, through Childress, Sweetwater, San Angelo, Fort Stockton, Alpine and Presidio, Texas, and then on to Topolobampo via Chihuahua City.

He organized the Kansas City, Mexico & Orient Railroad, still incorporated, and started survey work as far back as 1904. When he folded up, there were large gaps, both in the United States and Mexico. The Santa Fe system became the owner of the KCM&O, and a branch of the Mexican government later took over the Mexican section.

FINISHED IN 1930

The Santa Fe finished the stretch between Alpine and Presidio Nov. 2, 1930, when 200 Texans, riding a special train to Chihuahua City, watched Señor Luís L. León, personal representative of President Ortiz Rubio, drive a silver spike.

In Mexico, the railroad was built from Ojinaga at the Rio Grande to Chihuahua City, then on to Creel[1], west of the Continental Divide and 180 miles west of Chihuahua.

Reports in 1930 were to the effect Mexico was going to complete the railroad in a short time, but Mexico also had its depression and it seemed as if the Stilwell dream would never materialize.

However, Mexican business men and engineers, especially the three Togno brothers, began to realize this section of Mexico had to have a railroad before it could be developed. A section of mountainous country as large as Louisiana was devoid of both highways and railroads and in this area are tremendous natural resources.

Survey crews were sent into the Sierra Madre, between San Pedro, Sinaloa, on the west and Creel, Chihuahua on the east. From their findings the Department of Communications and Public Works, known as SCOP, made and submitted plans to President Ruíz Cortines, who now is committed to completion of the Chihuahua-Topolobampo railroad within three years.

On the unfinished eastern section, between Creel and Témoris, 93 miles, there are three modern camps for the SCOP engineers. Seven small con-

1 The railroad reached Creel in 1907.

The Mennonites

In 1874, 10,000 Mennonites left Russia for the USA and 8,000 for Canada. Then between 1922 and 1926, 6,000 of the Canadian Mennonites moved to Cuauhtémoc and, in 1924, 1,000 moved from Canada to the state of Durango.

They came on 36 chartered trains with everything they would need, including animals. The transition was not easy. They had to learn how to make adobes for building houses, and some of the crops they were used to planting in Canada did not do well in Mexico.

And the trip was, in fact, scary. The Mexican Revolution had only officially ended two years before and Pancho Villa was still alive. Mennonites have told me how their grandparents were scared crossing into Mexico at Juárez because the Mexicans (people of a different color and speaking a different language) would jump up, hanging onto the windows of the train just to see these strange white people.

And when the train left Chihuahua City, in order to get up the steep grade towards Cuauhtémoc, the train engineers had to uncouple some of the cars and go back for them later. Not understanding what was going on, some of the Mennonites were really scared. One Mennonite told me that his grandmother, who was with that first bunch, was blind. We can only imagine what was going on in her mind.

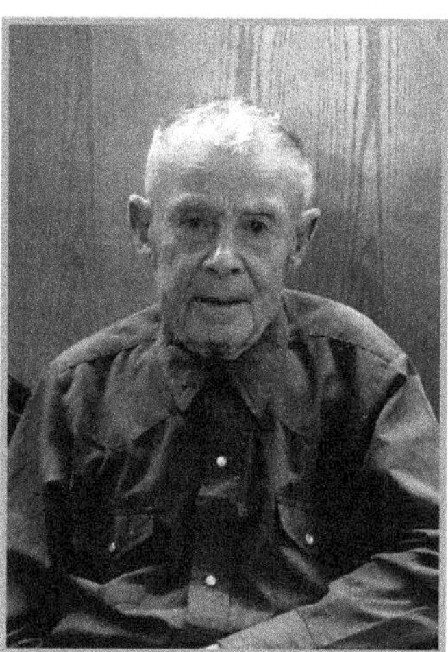

4.11. Abram Thiessen was 4 years old when he arrived in Cuauhtémoc in 1922. Only one other man (Cornelio Rempel) and several women from that first group of immigrants are still living as of 2012 (photo D.B. 2012).

Today there are approximately 40,000 Mennonites in Chihuahua and 40,000 others in seven Mexican states. Others have taken up residence in several other Latin American countries and in places like Seminole, Texas.

The people who work at the Mennonite Museum near Cuauhtémoc, such as Lisa Wolf and Enrique Wolf, are very knowledgeable and helpful. Also, see Gary S. Elbow's "Mennonites," *Handbook of Texas*, *Puerta a la sierra: Recuento historico de Cuauhtémoc* by Victoriano Díaz Gutiérrez (cronista de la ciudad), and *Ein feste Burg ist unser Gott (Der Wanderweg eines christlichen Siedlervolkes)* by Walter Schmiedehaus.

tractors are at work. On the western end, from San Pedro to Las Guazas on the Chínipas River, the Águila Construction Company, largest in Mexico, is working on 50 miles of roadbed. There are three SCOP camps on the western side. The Águila Company is completing 1.55 miles per month of grading and drainage.

JEEP ROADS ON ROUTE

Jeep roads follow the route with the exception of the Septentrión Canyon, just east of Las Guazas. This canyon is 21 miles in length and the Águila people plan to have a truck and jeep road through it within 12 months.

The general route from Chihuahua runs southwest and follows the drainage pattern of the western slopes of the Sierra Madre. From Chihuahua, the railroad breaks through some low mountains into the Valley of Cuauhtémoc, where dry land farming is practiced. Just outside of this town is a rather large colony of Mennonites who moved from Canada to Cuauhtémoc. Their homes, windmills and farms resemble an early German settlement on the Dakota prairies.

A few miles east of Cuauhtémoc a large cellulose, or pulp mill, is nearing completion. It will begin operation in late fall, is of Italian design and will use small pine trees and branches not suitable for lumber. It will draw its material from the Creel district and is closely allied with Maderas de Chihuahua, an old established lumber company.

Roads in the Sierra Madre

Just a comment on the roads, from the article I wrote for the *Alpine Avalanche*, Aug. 20, 1959, after spending six weeks working in the Sierra:

> During the rainy season transportation becomes a problem because the roads are quite often washed away or covered over by landslides. For instance, once we had to leave our truck because of a large landslide and walk several miles to where a Jeep was located. This Jeep was without a windshield, lights for the tunnels or brakes, and we had to use it to climb out of a 3000 ft. deep canyon. Whenever I wanted to take a picture, we could not stop so I would jump out, take my picture, and then run and catch up with the Jeep. At one place, we passed a rock 70 ft. high which had slid down into the grade.

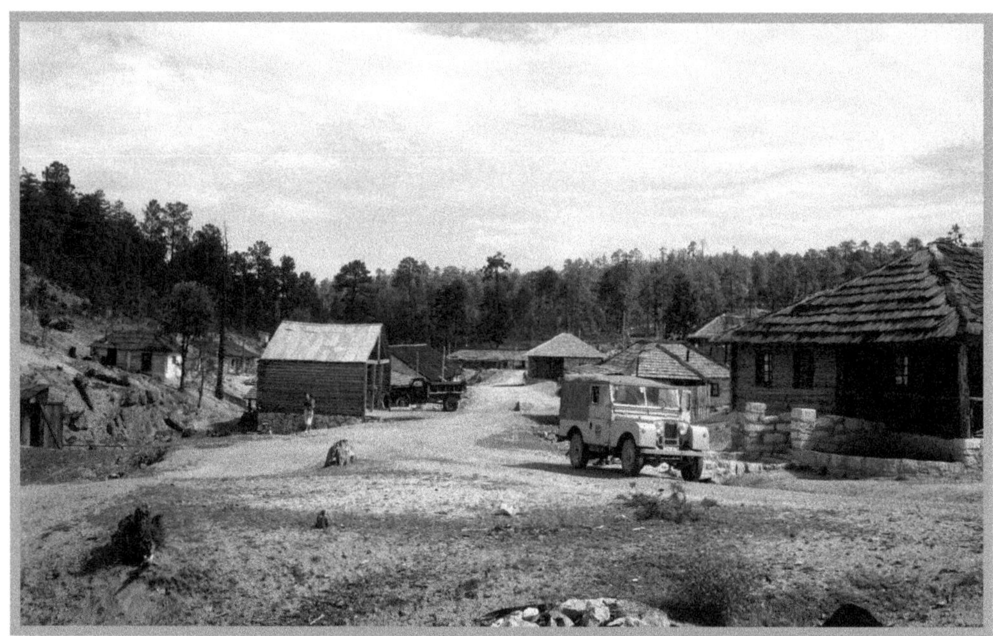

4.12. **CAMP FOR ENGINEERS**—Camp Areponápuchi, west of Creel, in the timber country, has been established 10 years. Engineers working on the new rail line to the West Coast are housed here (photo G.B. 1955). [Engineers referred to this camp as the University of Arepo because so many young engineers learned about topography from Francisco Togno at this place. The name Areponápuchi is linguistically 3 Tarahumara words formed into one: alé 'there', epó 'flat place', and napúchi 'a pass'.]

4.13. **NEW TOWN NUCLEUS**—A community called Anáhuac, to house the families of 2,500 workmen who will be employed in it will be built around the pulp mill being constructed on the new Chihuahua to Pacific Railroad, near Cuauhtémoc, about 40 miles west of Chihuahua City (photo G.B. 1955).

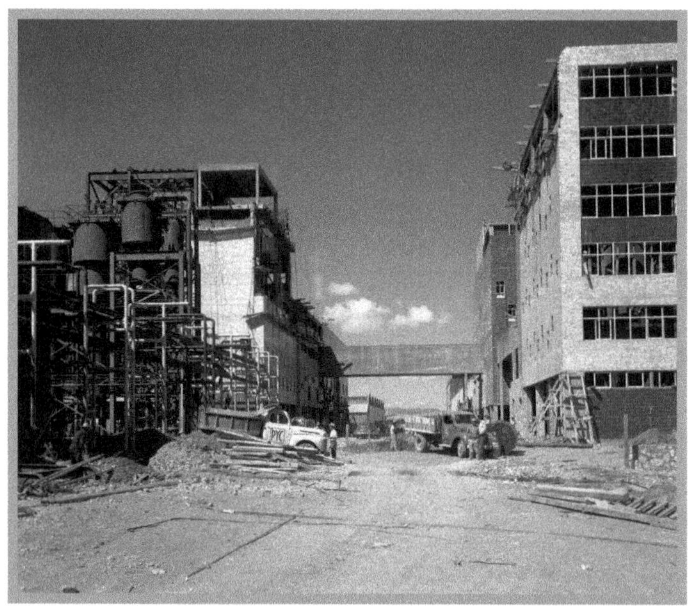

The mill is located in a valley that resembles the Marfa, Texas area [that is, high plains] and will be a complete community within itself. Twelve hundred persons will be employed.

West from Cuauhtémoc, the railroad crosses the Continental Divide at a rather low altitude and makes its way through the wheat, corn and apple orchards of the high valley with La Junta as its center. At La Junta, a branch of the railroad goes north to Casas Grandes, then on to Ciudad Juárez.

Thirty miles west of La Junta, the railroad begins to enter the timber country with its ponderosa, royal pine and fine oak. Five miles farther, it again crosses the Continental Divide and enters San Juanito, population 2,300, elevation 7,937 feet. The rainfall here drains into the Rio Conchos, that joins the Rio Grande at Presidio and Ojinaga[2].

THE JEMEZ MOUNTAINS

The San Juanito area is similar to the Jemez Mountain country around Los Alamos, New Mexico. It is the beginning of Tarahumara Indian country.[3] Most of its income is derived from lumbering.

About 24 miles southwest of San Juanito and on the Pacific side of the Continental Divide is Creel, the end of the line. Its population is 1,500 and the altitude is 7,900 feet.[4] It is a lumbering and mining town. From here it is a short trip to the Barranca del Cobre on the Urique River.

A scenic road, 120 miles in length, crosses two upper canyons of the Barranca del Cobre to reach the La Bufa Mine on the Rio Batopilas, which must be one of the most fantastic locations for a mine in North America.

From Creel, the early surveyors for Stilwell took off on foot and on mule and horseback toward the port of Topolobampo.

All the 175 miles of unfinished railroad is in forested mountain country. The completed railroad will be breathtaking every foot of the way. Cuts 100 feet deep, tunnels, one more than a mile long, shelf grades clinging to perpendicular mountain sides and

2 San Juanito was founded as a railroad town in 1903 and named for a local ranch. The first rock building built by the railroad, located by the tracks near the center of town, is still in use. See Carlos Jaime's *San Juanito, Crónica de lo que el tiempo dejó.*

3 See "Effects of the Railroad on Local Cultures" at the end of this book.

4 The geographer Dr. Robert Schmidt gives the altitude as 7640 feet. See his map (International Map Co. UTEP, Box 400, El Paso, TX 79968).

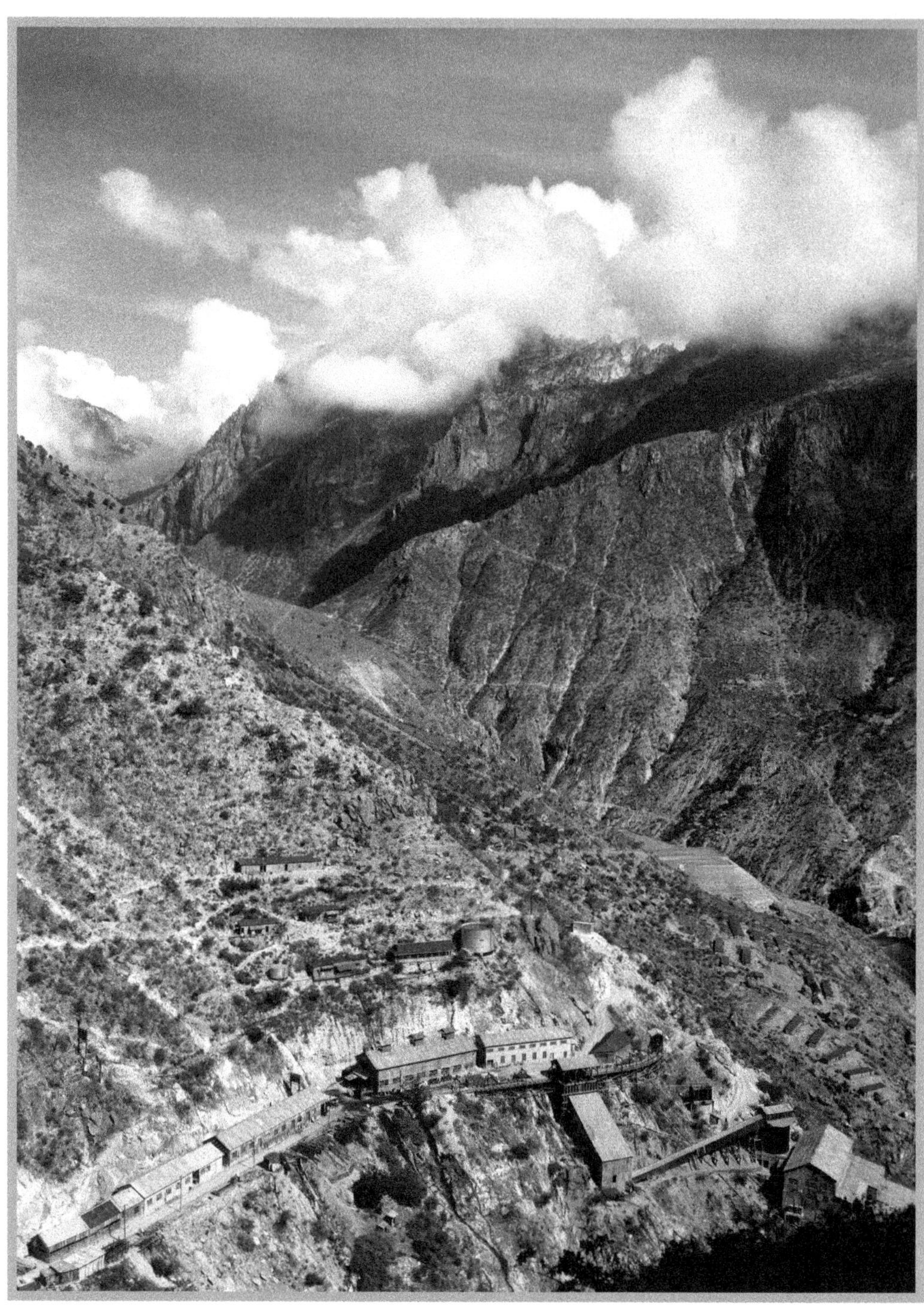

4.14. The copper mine at La Bufa, taken from 'La Noria', where engineers lived (photo G.B. 1955).

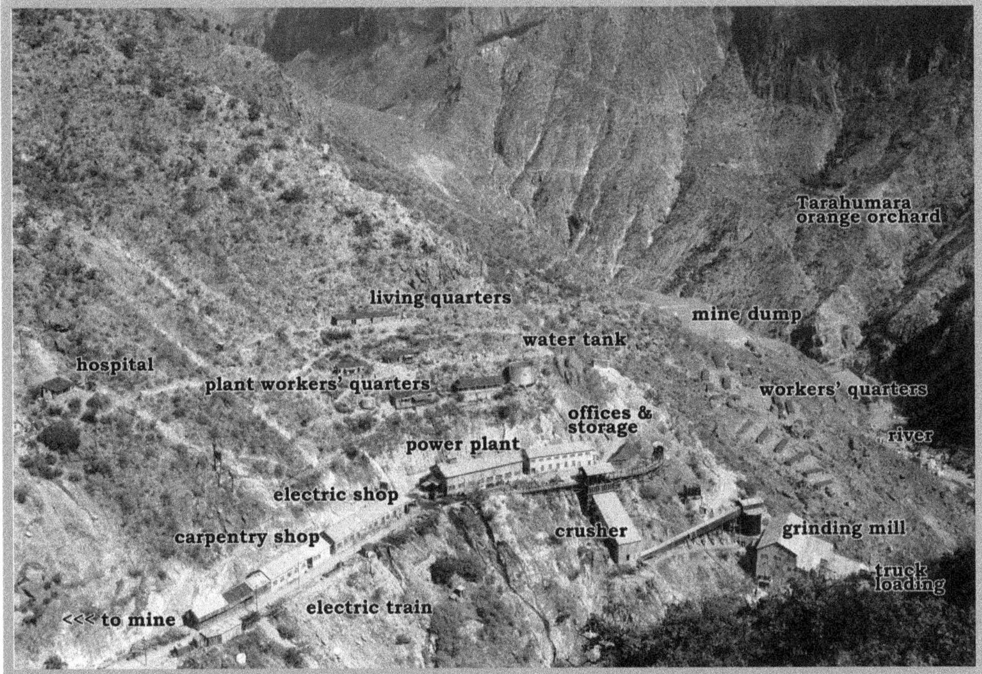

4.15. The copper mine at La Bufa with labels (photo G.B. 1955).

The Road to La Bufa Copper Mine

Glenn wrote, "A scenic road, 120 miles in length, crosses two upper canyons of the Barranca del Cobre to reach the La Bufa Mine on the Rio Batopilas, which must be one of the most fantastic locations for a mine in North America."

José Gándara of the Pemex Travel Department wrote a description of the area for Glenn. He said of the road descending to the mine: "The road is well traced, easy on the tires but hard on the nervous system, for while it is very smooth, it is narrow and snakes along the edge of precipices, which make it one of the most exciting trips one can make."

Erle Stanley Gardner put it like this:

> What makes the road so frightening is the realization that if anything goes wrong a man is completely helpless.
>
> A board three feet in width is a boulevard if it is raised only two feet above the ground. You could skip a rope on it, walk, run or lie down.
>
> But take that same board three feet in width and put it across a sheer chasm of five thousand feet and you would be crawling across the board, clinging to the edges, and covered with perspiration by the time you reached the other side—provided you didn't fall off en route (p.245).

The road is now being paved.

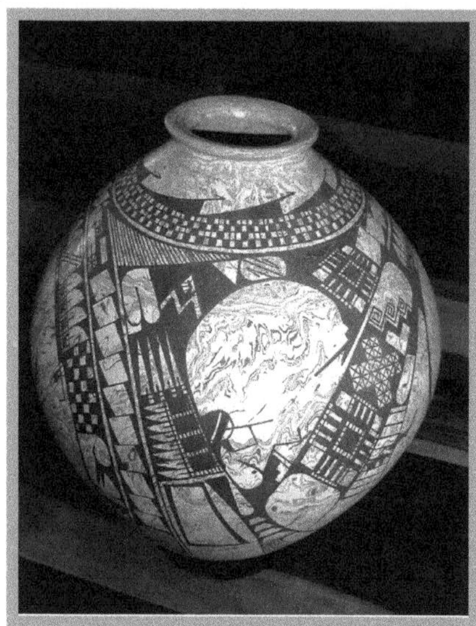

4.16. Tarahumara playing the flute design: Made by Mata Ortiz potters, Chevo Ortega Moreno and wife (He makes the pottery and she paints the original designs.) (photó D.B. 2012).

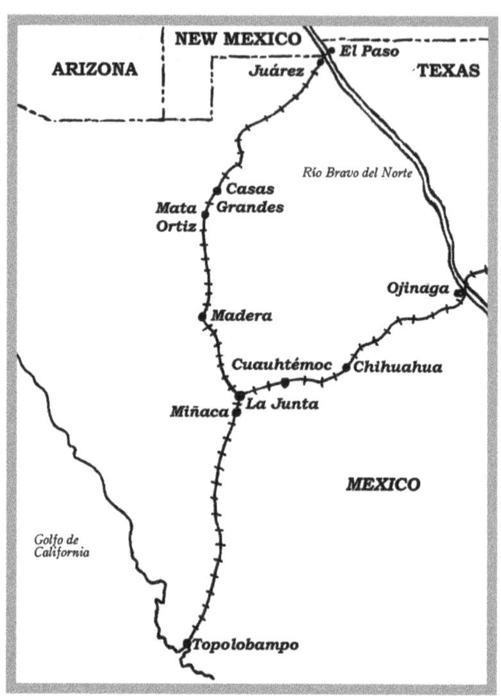

4.17. Sketch of the Noroeste Railroad (Juárez--La Junta) and the Chihuahua al Pacífico.

The Noroeste de México Rail Line

This line, the Noroeste de México, was completed in 1909 (See Almada, *El Ferrocarril de Chihuahua al Pacífico*, pp.156-157). It connected Chihuahua City with Juárez via some of the lumbering and mining areas of Western Chihuahua.

The history of the line is very interesting from the standpoint of the role it played in the Mexican Revolution, such as when "bandits set fire to a railway tunnel...shortly before a passenger train entered it, and blew up both exists... so the passengers were either burned to death or suffocated"(Katz p.415). It is also of interest from the standpoint of the colorful figures who were involved in the railroad, such as the owner of the Cananea copper mines, Colonel William C. Greene (See C.L. Sonnichsen's *Colonel Greene and the Copper Skyrocket*), Senator George Hearst, father of William Randolf Hurst, who owned a huge ranch in that area (See J.L. Kerr's *Destination Topolobampo*, pp.107-119), and the American electrical engineer and entrepreneur Frederick Stark Pearson, who owned a large lumber mill in that area.

The town that grew up around the mill was named Pearson, which today is the pottery making town of Mata Ortiz. Pearson and his wife died in the Lusitania, sunk by a German submarine in World War I (See *Journal of the Southwest*, Vol. 54, #1, 2012).

I rode on this line when I went to work on the construction in 1959, leaving from Juárez. The train stopped for the night in Madera, where the passengers all slept on the floor of the station.

plunging water courses to be crossed dominate the route. In this 175 miles, the railroad will drop from 8,118 feet to almost sea level, from high elevation mountain forests to tropical vegetation.

At one point, only 36 miles southwest of Creel, the passenger will find himself surrounded by pine forests, and if in January, possibly snow. From his train window, not more than three miles away and a mile below, he may see orange and avocado trees.

So important will be this route from a tourist standpoint, Francisco Togno[5] predicts streamlined trains will run from Chicago, across West Texas to Chihuahua, then over the mountains to Mazatlán.

5 Ing. Togno was chief of railroad construction in Mexico at this time.

Tarahumaras and the La Bufa Copper Mine Golf Course

Up on the top of the canyon, the miners made a golf course. After watching the game for a time, some Tarahumaras started making their own clubs out of oak branches, playing the miners and beating them. A couple of the Tarahumara players were taken to Juárez to play in a tournament, but one of them was hurt when jumping off the high board of a swimming pool. I saw one of the clubs, very well made, in El Paso at the home of anthropologist Robert Zingg's widow. Zingg and his partner Wendell Bennett spent a year in the Tarahumara (Samachique) area in the 1930s.

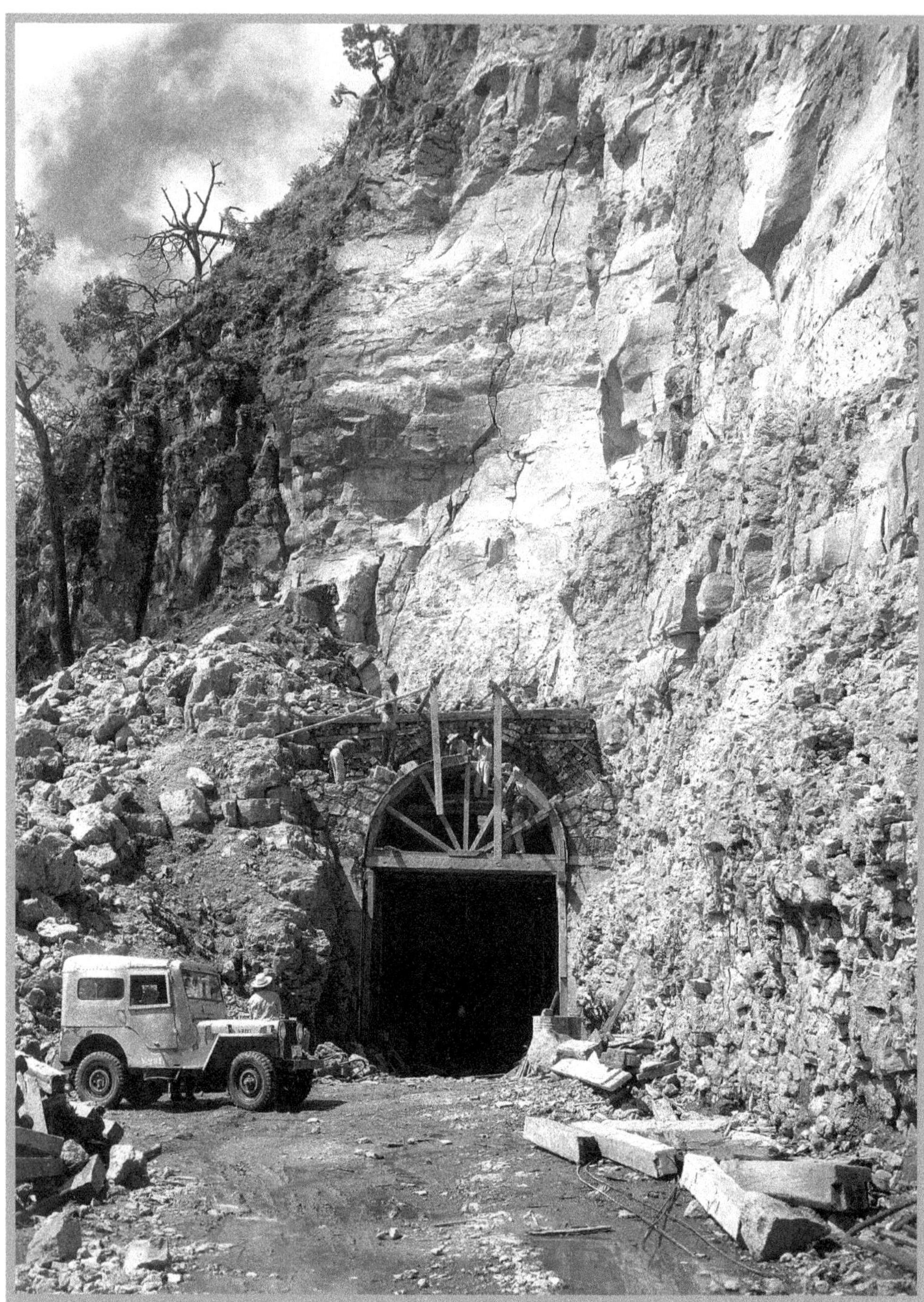

4.18. **TUNNELS FREQUENT**—Many tunnels are being driven through rugged mountains along the Chihuahua to Pacific Railroad. Few are straight. Most are self supporting, but this one will have concrete and stone support (photo G.B. 1955).

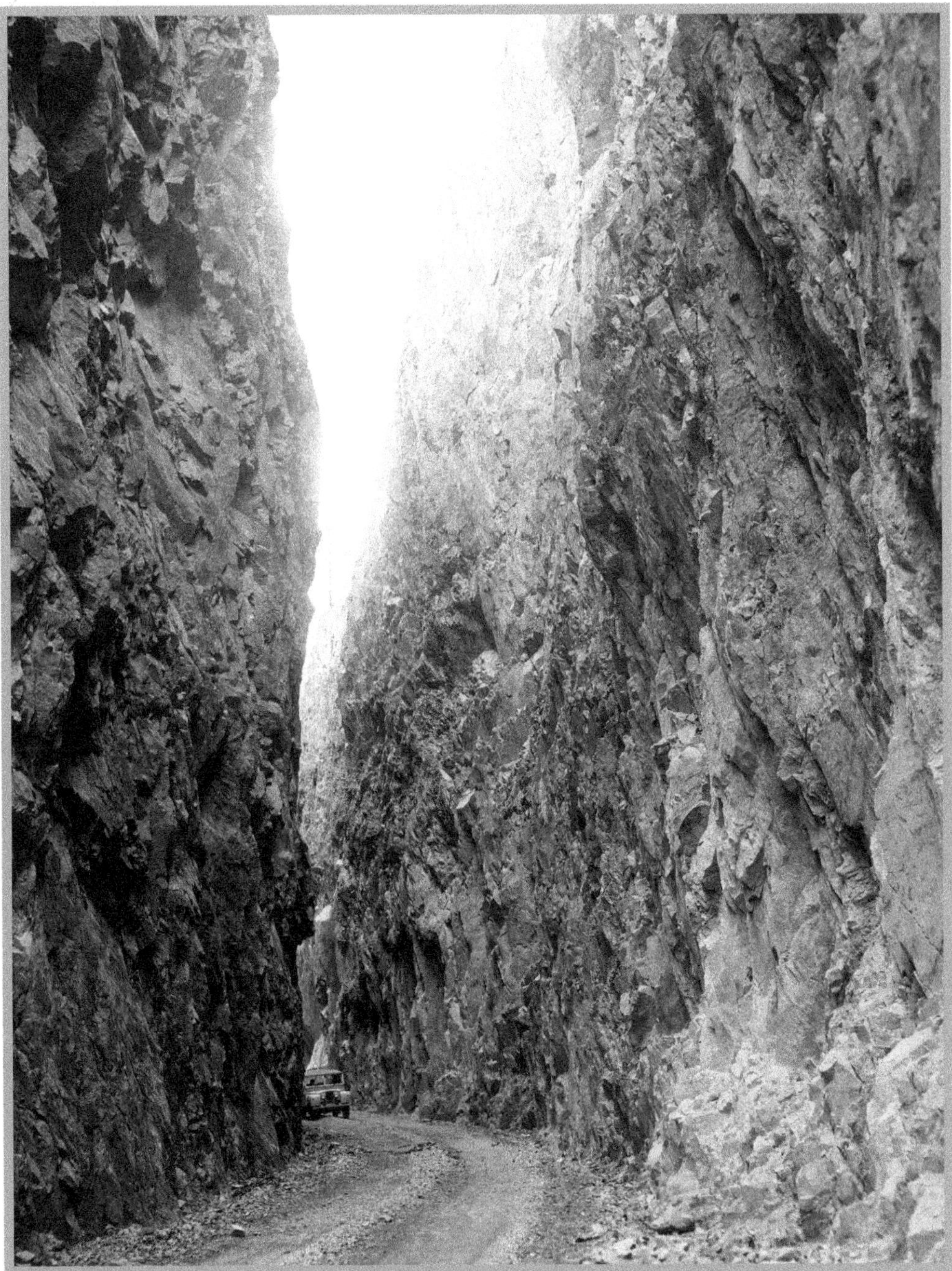

4.19. **DEEP CUT FOR RAILWAY**—The track of the Chihuahua to Pacific Railroad will be laid through this deep cut, made by American construction crews many years ago, near Creel (photo G.B. 1955). [This cut, known as "Cuatro Vientos" (four winds) because of the strong winds that whipped through it, was mostly made using dynamite, pick, and shovel. Today it has a "false" tunnel roof built over it to keep rock from falling on the track.]

Possible Routes for the Chihuahua al Pacífico Railroad

Numerous routes were studied through the years. Ing. José Eloy Yáñez lists eleven that Francisco Togno had to decide between ("Apuntes Históricos de las rutas...."). Several of these involved going down the Oteros Canyon, which becomes the Chínipas Canyon. Another involved going down into the Urique Canyon with a cog-wheel system to get the train up and down the steep inclines. The final choice was to go down the Septentrión Canyon. Owen made a survey trip in 1887. Ulisis Irigoyen, along with others, made another survey trip in 1938, and in 1939 the historian and governor of Chihuahua, Francisco Almada, who was from Chínipas, accompanied some engineers on a survey through the mountains (Almada, *El Ferrocarril...* pp.137-139). The Togno brothers, Francisco, Jorge, Ramón and others spent numerous months in 1941 trekking these canyons trying to find the best route. Kerr mentions a reconnaissance survey made by one of the Santa Fe Railroad engineers in 1928 (p.173). A clue as to why the Septentrión Canyon was selected for the route can be found in engineer Fred Fitch's letter in 1881 (*The Texas, Topolobampo & Pacific Railroad and Telegraph Company: Reports*, p.43):

> The Sierra Madre range of mountains, which traverses Mexico, is, as a general thing very much broken,--irregular tangents and curves, cut up by deep gorges and ravines, which assume no regular course or direction. A notable and very remarkable exception to this fact is the Setentrion [sic.] ravine or gulch, which cuts this lofty range of mountains through from their summit to their base. This gorge runs a general course of N. 30 degrees E. for a distance of 93 miles, little more or less, and forms the only feasible natural opening through the mountain range which offers an easy grade to overcome the 6,000 feet, more or less, which the track must reach.

Glenn wrote in his thesis concerning the survey work of the Togno brothers: "The mapping and engineering work had to be done without benefit of aerial or contour maps. No elevation points had been established. For many months Ing. Togno and his young wife lived in the back country. After he had set up a string of engineering camps, he made the trip from Creel to Choix, a village near El Fuerte, every two months by horseback, a round-trip distance of 352 miles. It took twelve days for him to make the trip."

One lady, the owner of a large, remote, citrus orchard at Huachajuri, not far from the Chínipas River, told me that the orchard was put in because of the hope that the railroad would go near there, but it did not. In the end, the shortest was selected, one that had been suggested by Owen back in the late 1800s, the Septentrión Canyon.

Branch lines were also discussed throughout the years, such as lines to the mining towns of Batopilas and Urique, which, if attempted, would have been very exciting engineering feats. A branch line was completed from Cuauhtémoc to the mine of Cusihuiriachi in 1911, which, according to Díaz Gutiérrez, was part of the Noroeste Railroad, which went from La Junta to Casas Grandes and Juárez (pp.21-22).

The El Paso Times

– Tuesday, September 24, 1957

Railroad To Unite Sinaloa, EP, Presidio

by Glenn Burgess
Times Correspondent

Topolobampo, Sinaloa, Mexico—In many ways, the Mexican State of Sinaloa represents the "pot of gold" at the end of the rainbow for business men of the El Paso and Presidio areas. Especially will this be true one year from November when the Chihuahua al Pacífico Railway is scheduled to begin the operation of trains between El Paso and Presidio, and Topolobampo on the coast of the Gulf of California.[1] For fishermen and tourists, Mazatlán in the southwest corner of Sinaloa will be the final answer for many.

Sinaloa is rapidly becoming the center of one of the richest agricultural empires of North America. The Port of Topolobampo will be the nearest Pacific port to mid-continental United States, while Mazatlán is scheduled to become both the commercial and tourist metropolis of the west coast of all Mexico. Topolobampo is nearer to Presidio than Austin, Waco and Fort Worth, and Mazatlán is the same airline distance as Corpus Christi and Waco. The completion of the Chihuahua al Pacífico Railroad will make this magic region a part of the economy of El Paso, Presidio, Alpine and points to the northeast.

To reach the agricultural regions of Sinaloa, the railroad, begun from Kansas City in 1900, is having to cross the mighty Sierra Madre,[2] breaking out through the Septentrión Canyon on the border of Chihuahua and Sinaloa. Through this 21-mile-long canyon the railroad will lose elevation swiftly, then 20 miles of semi-tropical foothills brings the road to the first of the irrigable plains of Sinaloa. Just after leaving the Chínipas River gorge, the rails will have to go through a

1 It was not finished until 1961.

2 Note that this is not the Sierra Madre made famous by the Humphrey Bogart movie "Treasure of the Sierra Madre." That was about the Eastern Sierra Madre, not the Western.

tunnel 5,051 feet long.[3]

RUNS ALONG RIVERS

From Creel, the present end of the Chihuahua line, the railroad parallels the main streams of the Fuerte River system—the first the Urique River and its deep Barranca del Cobre, then the Fuerte itself. The Rio Fuerte actually leaves the mountains at the Cañon de Huites where the Mexican Government plans a 600-foot hydroelectric dam.[4] A short distance below this canyon, the railroad will cross the Rio Fuerte over a bridge 480 meters long and 50 high. It is only a few miles from this bridge to the present rail head of Sinaloa, [the town of] San Pedro. A train now runs over the 77.6 miles between San Pedro and Topolobampo.

The Mahone or San Miguel Dam at San Pedro is the most important key to Sinaloa's growing agricultural prosperity. The dam was completed in February, 1956. It is an earthen dam 216 feet high and 1.86 miles long. The project cost 160 million pesos. It opens up 500,000 acres of new irrigated land that will support 310,000 additional people.

If all this new land were put in cotton, it would produce as much as the entire Texas South Plains block of cotton counties.

TO BUILD CANALS

Between the dam and the Gulf of California is an extremely rich and level flood plain with just enough fall to drain the deep soil. The plain now receives insufficient rainfall for cropping, however, and is covered with thorn bush and cactus. The cost of clearing and repairing this land runs approximately $22.40 per acre. To distribute the water, there will be 100 miles of major canals and 124 miles of lateral canals.

Original plans called for 70 per cent of the farming operation to be mechanized. Settlers are coming from all over Mexico and even a few from Texas and New Mexico. The state of Sinaloa is supplying around 80 per cent of the new farmers.

Before the advent of the huge dam, the Fuerte River Valley produced cotton, sugar cane, wheat, tomatoes, chick peas, and corn. The tomato crop matures earlier than any in the United States and commands

3 The Ferromex office gives the length as 6,024 feet or 1836 meters. Schmidt's latest map puts it at 5,966 feet or 1820 meters. According to Ing. Leal, this tunnel had to be built because of plans to build the Huites Dam which would back water into the areas that otherwise could have been used for the railroad.

4 The Huites Dam was finished in 1995.

premium prices. Fruits and other vegetables also are grown in this semi-tropical climate.

The town of Los Mochis, on the Chihuahua al Pacífico and 12 miles from Topolobampo, is benefitting most from the new dam. It was established in 1902 and had 30,000 inhabitants in 1955. Its wide streets and activities resemble those of earlier towns in the Texas Panhandle and on the South Plains.

BIG FARMING REGION

The Chihuahua al Pacífico's agricultural region stretches from Ciudad Obregón, 150 miles north of Los Mochis to Culiacán, capital city of Sinaloa, 124 miles south. The Fuerte River is one of 12 rivers that plunge out of the wet and scenic Sierra Madres, then cross Sinaloa's fertile flood plains to the Gulf of California. The Tamazula River at Culiacán has the Sanalona Dam that impounds water for 140,500,000 pesos worth of crops. Before the Mahone Dam, the Fuerte River Valley produced 145,107,000 pesos worth of crops. The 10 remaining rivers of Sinaloa await the construction of dams.

Just what will be the value of the new production of the Mahone Dam district remains to be seen, but it will run into millions of dollars per year.

Mazatlán has a head start in Sinaloa on importing cotton and other agricultural products. The town is so important that officials of the Chihuahua al Pacífico have the dream of running a vista-dome-type train from Chicago to San Blas (near Topolobampo), to Mazatlán and entering Mexico at Presidio and Ojinaga. Freighters from all over the world call at the Mazatlán Port. Fishermen from California, Arizona, New Mexico and Texas believe that from Topolobampo to Mazatlán is the continent's finest fishing grounds.

The climate of Mazatlán is considerably more equitable than that of Acapulco. Summer temperatures run from 70 to 85 degrees, and the winters 70-75. Sixty-eight degrees is the minimum recorded and 88 the maximum. As one hotel man who is building a 400-suite hotel remarked—Mazatlán is the place where people of Florida come to spend the winter.

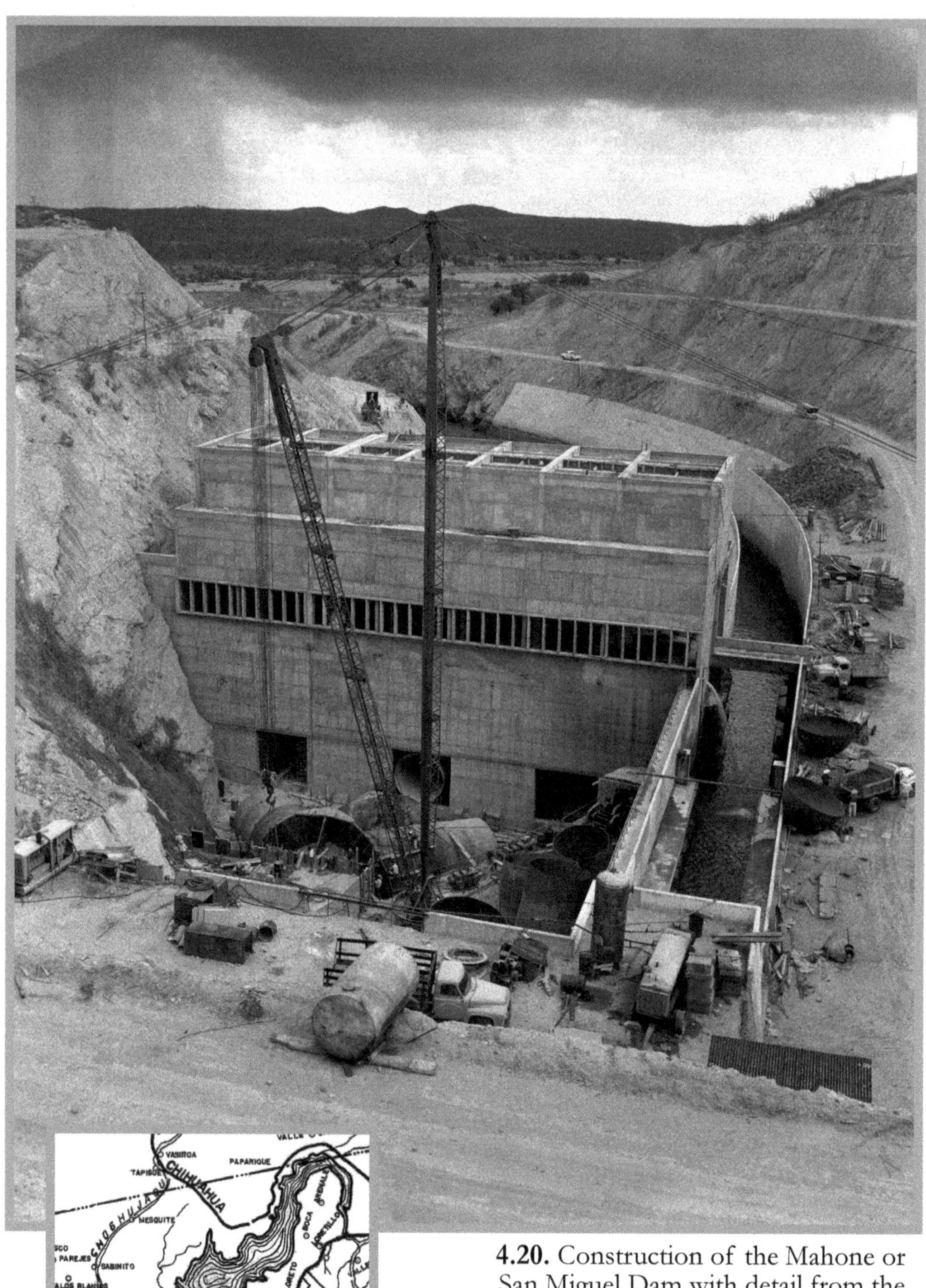

4.20. Construction of the Mahone or San Miguel Dam with detail from the 1953 SCOP map (photo G.B. 1955).

CLEAN MARKET

The health of Mazatlán people is good. The swamps have been drained, the mosquitoes eradicated, and its large city market is probably the cleanest in Mexico.

With the exception of shrimp fishing, Topolobampo's future lies ahead. It claims the largest freezing plant for shrimp in the world. The population varies from 250 to 5,000 depending on the fishing season—the peak is in December. The packing plant is now freezing halibut and other deep sea fish during the summer or off season to maintain skeleton crews.

The natural harbor of Topolobampo is 11 miles long, and averages 270 feet deep. All the fleets of the United States could be anchored there. This harbor, with completion of the Chihuahua al Pacífico in 1958[5], will give mid-Continental U.S. access to the ports of the South Pacific, Asia, and the west coasts of Mexico and South America.

5 It was finished in 1961.

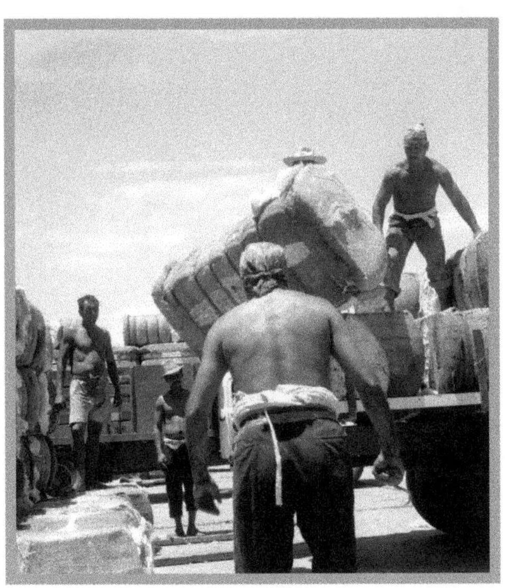

4.21. Unloading cotton bales in Mazatlán (photo G.B. 1955).

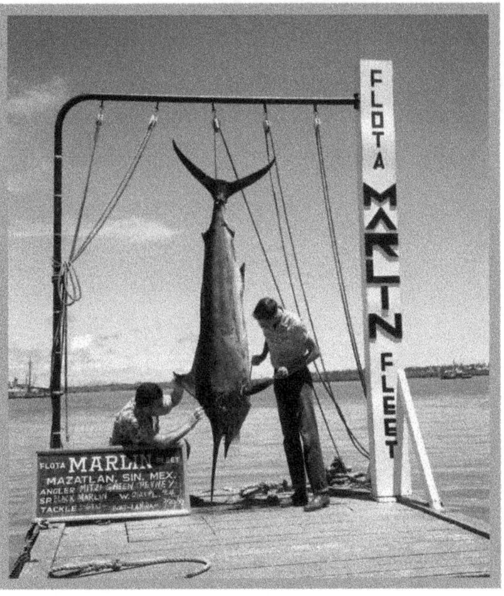

4.22. Two hundred pound Marlin at Mazatlán (photo G.B. 1955).

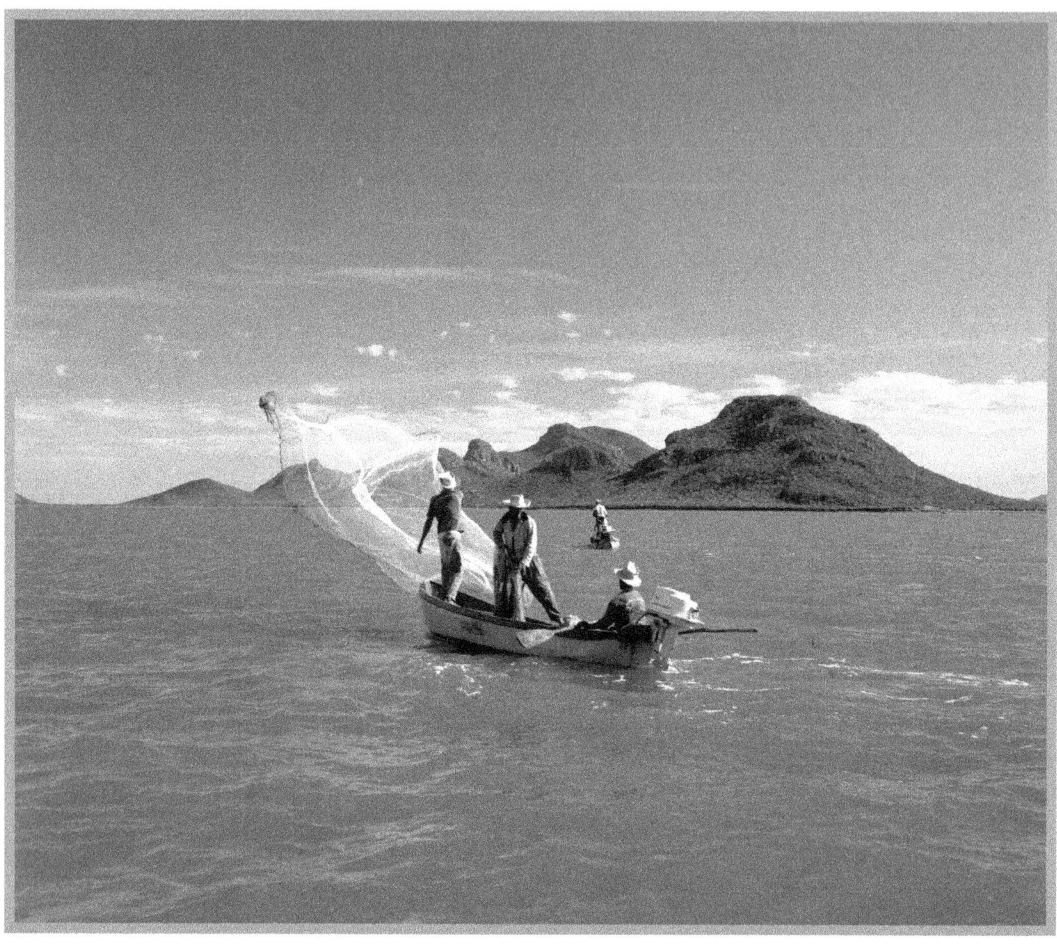

4.23. Fishermen throwing net on the Bay of Topolobampo (photo G.B. 1955).

The Bay of Topolobampo

In the report to Owen in 1881, George Simmons described the Bay of Topolobampo by saying: "I am familiar with many of the finest harbors in the world, but for natural beauty I know of none that excel, and few that equal, the Bay of Topolobampo. We dismounted from our mules and sat for some time upon the shore, admiring the graceful lines of the harbor, watching the sea-fowl, and speculating upon the time when this spot should be alive with the business of all nations" (p.7).

The El Paso Times

– Saturday, May 2, 1959

Chihuahua Railroad Forges Pacific Link

by Glenn Burgess

Times Correspondent[1]

Chihuahua, Chih., Mexico.—For fifty years, economists and engineers have been dreaming and working toward a short route by rail to the Pacific from mid-continental America.[2]

Sometimes it has been men working on the development of the fertile agricultural empire known as the "bread-basket" of Mexico—southern Sonora and Sinaloa. Again, it was promoters from Kansas City, then back again to a capitalist of Sinaloa.

Finally after revolutions, and after the Santa Fe System had the opportunity of completing the route, but sold it for a song, the Government of Mexico is completing the project on the theory that the old Kansas City, Mexico & Orient, is more than a railroad.

Under its new name, the Chihuahua al Pacífico, it is a means of connecting two isolated regions that are separated by a high and rugged mountain range that runs from the southern border of Arizona, 1,200 miles south before railroads and highways cross it near Guadalajara not far from Mexico City itself.

The new railroad is a development idea. It is reaching into regions rich in mineral, forest, and agricultural products that can be reached only on foot or by horse and mule. It will make possible the movement of products from the Pacific Ocean across Mexico to Fort Worth, Dallas, Oklahoma City, Kansas City, St. Louis, and Chicago, at a saving in the all important "freight rate".

This new line taps the new 500,000-acre farming region created as a result of the completion three years ago of the Miguel Hidalgo (Mahone) Dam across the mighty Rio Fuerte, the largest of Sinaloa's 12 rivers that rise in the rainy western Sierra Madre and rush across rich flood plains to the Gulf of California. It makes available the resources of the verdant

1 Another article "Dream Mexico Railroad Now Nears Completion", *Fort Worth Star-Telegram*, Sunday, May 10, 1959 is almost the same and is not included here.

2 Actually since 1849.

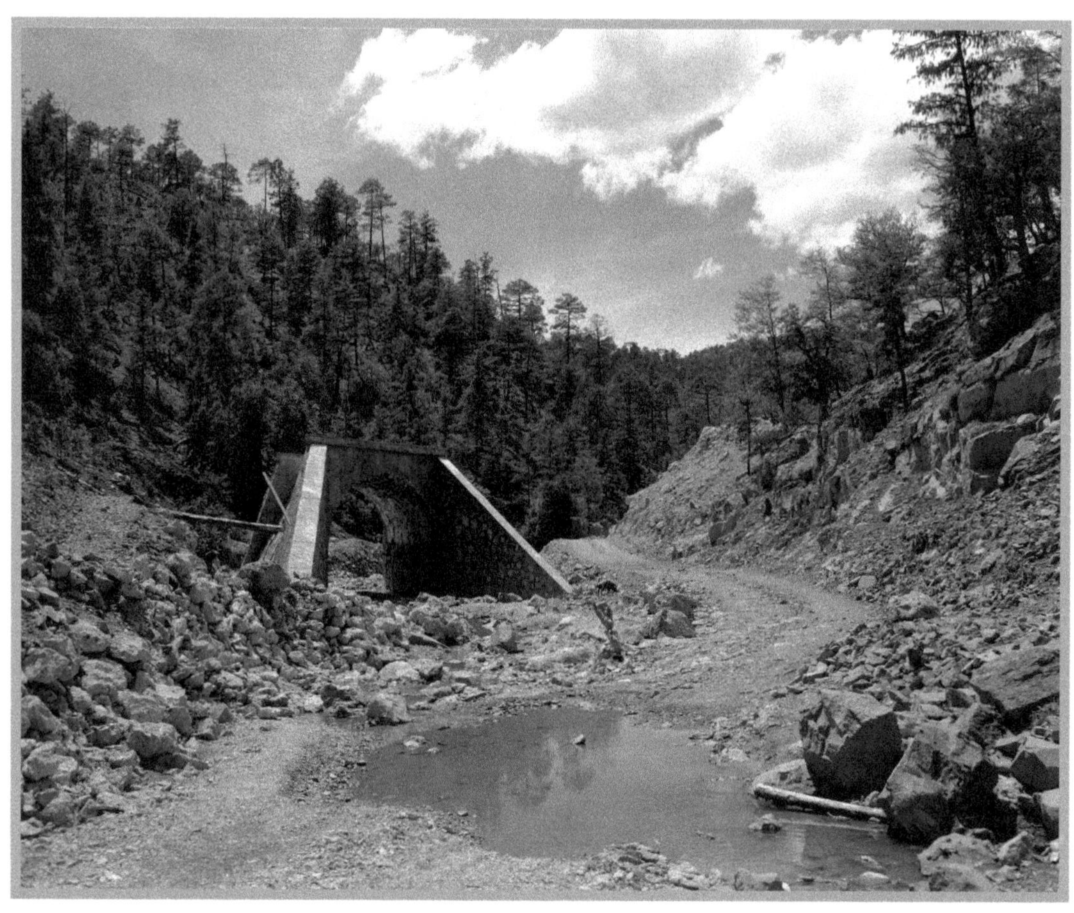

4.24. Culvert near Guasachique (photo G.B. 1955).

pine and oak forests that can come to the United States and to all of Mexico in the form of lumber, cross-ties and cellulose products.

As the railroad develops, it is expected that its branch lines will reach out in the form of highways and will open up to tourists one of the fantastic scenic and interesting regions of America until recently completely isolated.

That it will take three years longer to complete the grades, fills, tunnels and high bridges of which the line is a rapid succession, does not in any manner dampen the enthusiasm of those men who have spent practically a lifetime on the idea that might be termed the "Magnificent Obsession" of Mexico. While the era of railroad building in the United States was over years ago, here again, the railroad will come first, and following it will be the highway and airport development.[3]

Without the benefit of Mexican government funds, it is doubtful if the railroad started in 1900 by Arthur Stilwell of Kansas City, would ever have been completed, because private corporations were skeptical of the railroad's ability to make a return on the investment. Then, the government of Mexico nationalized the country's railroads and in 1940 took over the old Oriente. Now, the engineers working on the project believe that, with proper management, the railroad will pay its way, in spite of the staggering costs of construction and of rehabilitation of the old Oriente that has today 426 miles of rails.

Work on the project goes on in two phases. First, the entire line between Chihuahua City and Creel, 189 miles to the west, is under a rehabilitation program. This will also be done on the 70 miles in operation on the West Coast between the seaport of Topolobampo and San Pedro. Needed work will be done also on the 168 miles between Ojinaga and Presidio to Chihuahua City. When this modification work is complete, the entire line will be usable for all types of railroad cars and diesel locomotives and the movement of freight and passengers will be safer and much faster than has been the case with most railroads of Mexico.

The second phase that has interested southwest Texas and Mexico for several years is the construction through the rugged and high Sierra Madre of

3 As of 2012, a paved road across the mountains and an airport at Creel that can handle 100 passenger jets were nearing completion.

the remaining gap of 180 miles. The grades, fills, cuts, tunnels and bridges are 85 to 90 per cent complete on this 180 miles. Yet, the most difficult part remains, and with good steady work by large contractors, engineers now estimate it will be another three years before the rails can be laid.

The rehabilitation program has been unknown to most Texans. For a year or more, steel rails from the Colorado Fuel and Iron Co.'s mill in Pueblo, Colo., have been passing through San Angelo, Alpine, and Presidio, but it was not known how these were being used.[4]

Three years ago, Ing. Alfonso Rincón Benítez, Mexican engineer in charge of both the construction work and rehabilitation of the Chihuahua al Pacífico, had his plans ready for the remodeling job on the old line. In 1957, probably the top rehabilitation engineer of Mexico, the dynamic Ing. Mariano García Malo of Mexico City, was secured by Ing. Francisco Togno, Chief of Construction of Mexican Railways in the office of the Secretary of Communications and Public Works, to take over the difficult task of converting the old Chihuahua al Pacífico into a first class railroad line.

Grades have been made consistent, curves eliminated, at least two tunnels will result, and adequate bridges and culverts have been constructed.

First, the grade, after being revamped, receives a thin coating of asphalt, and on top of this comes good ballast crushed from volcanic lava rock. Next comes good crossties, most of them creosoted. Ballast and creosoted ties are fairly new for Mexican railroads.

The new rails between Chihuahua City and Creel are of 90 pound steel. For most of the distance three rails are welded together making single rails over 36 yards long.

On the three-rail system, they are introducing what is known as "fair rail anchors" to keep the rails from moving up or down the tracks, preventing warping and spreading.

Of the three-rail system, 151 miles is complete. This brings the engineers to an experiment in a European or French-Italian system. For 38 miles they are using rails welded together for a distance of 800 meters or one-half mile. Instead of the conventional fastenings, the ends of the rails are beveled and overlapping. Thus, the old time "clickety-clack" noise is

4 See Glenn's article "Sierra Challenge", The *CF&I Blast*, May 22, 1961.

almost completely eliminated

The long half-mile rails are set on rubber plates, and are fastened to cross ties with large screws and held by spring steel plates. This system absorbs the shock of the trains and cuts down the number of cross-ties needed by 15%. Also the spring rail anchors are not needed. Heavier traffic can move over this type of track.

On one short segment, they are experimenting with concrete cross ties made of two sections of concrete fastened together or connected by a short piece of old steel. It is estimated these will last 50 years as compared to the 30-year life of the average creosoted tie in the U.S..[5]

One of the tunnels planned is now under construction west of the small town of Bocoyna at the headwaters of the Rio Conchos and on the east side of the Continental Divide. It is 585 feet long and will cost $160,000. Another is planned a short distance farther west at Aguatos and will cost an estimated $640,000.

Between San Juanito, also on the headwaters of the Rio Conchos, and Bocoyna, was located a crusher plant. 100,000 cubic meters of ballast is now piled there. Another....[6]

Between Ojinaga and Chihuahua, much work has been done recently on improving this 168 mile section. Four bridges that were washed out by an arroyo that joins the Rio Conchos at Falomir are being replaced by substantial concrete bridges at a cost of $400,000 with completion scheduled for June 1.

Most of the Ojinaga-Chihuahua City route has 80 pound rails and the dirt and gravel grades are substantial. Bridges and culverts are of good construction. At present, modification work is not contemplated.

At La Junta, where the Chihuahua al Pacífico branches—one, the Noroeste goes north to Casas Grandes and Juárez, the other west to Creel and ultimately, the Gulf of California. A modern creosoting plant of German design is in full operation. A new roundhouse has also been built and is ready for machinery.[7]

Total length of the railroad

5 One of Stilwell's associates, W.W. Sylvester, reported in 1900 that the line would pass through "extensive, untouched forests of long leaf pine, oak, mahogany and a species of tree that makes railroad ties that are almost indestructible" (as quoted in Pletcher, p.276).

6 There is a printing error here that makes this sentence incomprehensible: Something about 70 miles west of Chihuahua.

7 This roundhouse was inaugurated in 1961 by President Adolfo López Mateos and was one of the largest in Mexico. It is no longer in operation.

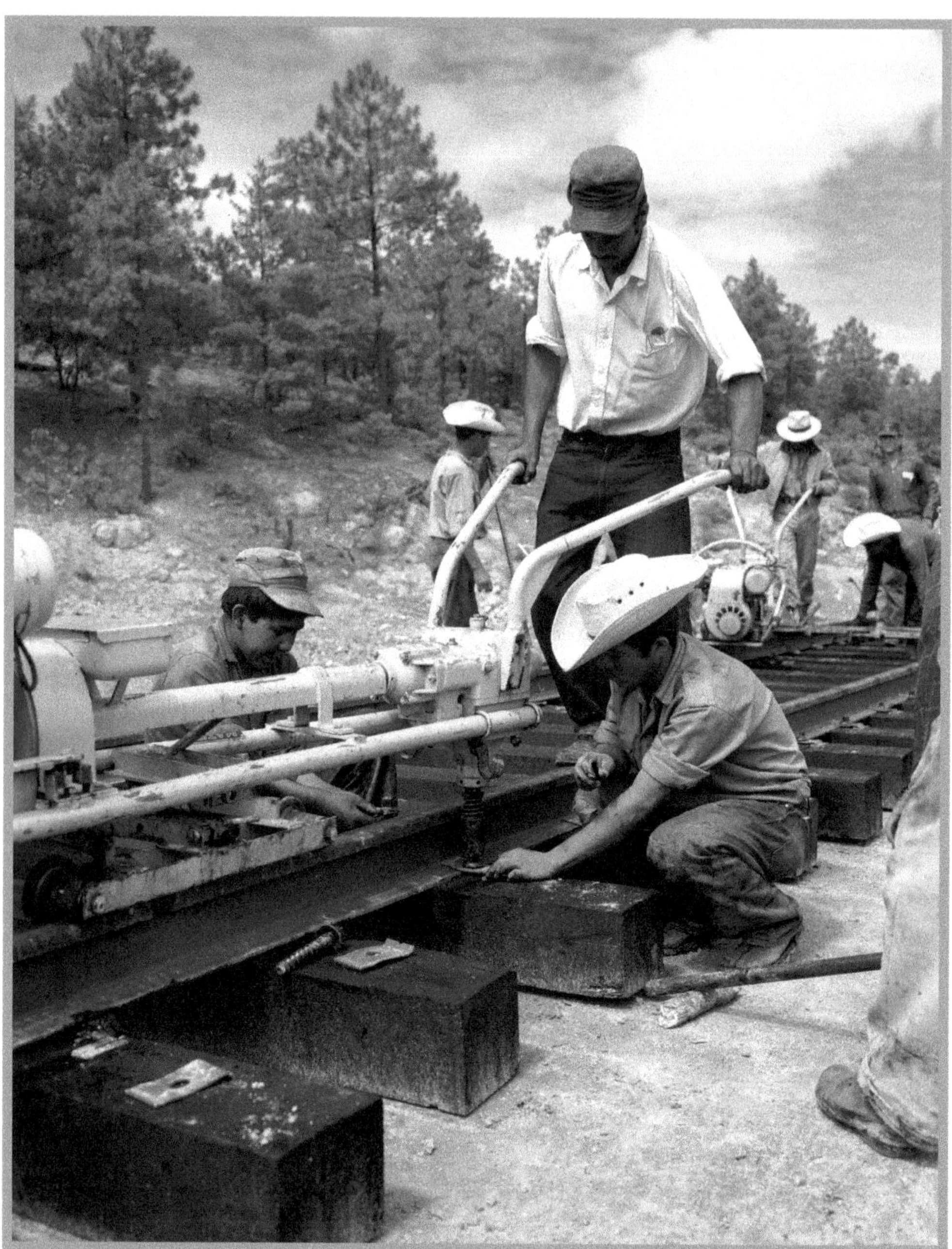

4.25. **LAYING STEEL**—The old spike driving gang has been replaced by a machine which twists a screw into a pre-drilled hole in a creosoted cross tie. The rails, set at an inward angle of 1:40, rest on rubber plates, and a spring plate is inserted between flange and screw (photo G.B. 1960).

from Ojinaga to Topolobampo on the west coast will be 606 miles. Rehabilitating the part that already has 426 miles of rails is hard, grinding work. Ing. Mariano García Malo is one engineer who can get the job done. Crowding 50 years, he has had a rich experience with both the Mexican National Railways and construction work under the Secretary of Communications and Public Works. His father was an engineer and professor of engineering before him. He is both a civil and a mechanical engineer. His practical training began in 1935 in Durango. Since then he has worked in Campeche, Puebla, Guadalajara and Guanajuato as Chief of Rehabilitation of the old Southern Pacific line between Guadalajara and Nogales, Arizona, and finally, as chief of rehabilitation of the Chihuahua al Pacífico.

García can read technical books in Spanish, English, Italian and French. He knows what he is doing, and wants those under him to do their work correctly. He lives railroad engineering and works long hours at it. This year he is in charge of construction work—rehabilitation—to the tune of $1,840,000. His department will need another $5,600,000 to put the Chihuahua al Pacífico into first class condition so that it will take its place among the top railroads of America.

However, no railroad in America will have the international significance of this one, and as important as the work of Mariano García Malo is, the pioneering challenge of the construction work going on in the western side of the mighty Sierra Madre of Chihuahua and Sinaloa must take the spotlight in North American railroad history. It is a project described by García, who went over this new construction route with the writer, as really a "work of the Romans."[8]

8 Schmidt says there are about eleven miles of tunnels and two and a quarter miles of bridges.

Working on a Spike Driving Gang

Talking recently with 71 year old Salvador "Chava" Bustillos, I learned that the screws were called 'tirafondos' and the plates 'galletas'. Chava, who worked as a young man on one of these crews, showed me one of his fingernails that was somewhat deformed. He had been holding one of the plates while the screw was being put in, and his finger got caught under the plate. Chava presently (2012) lives in Creel where his son is the "Presidente Seccional".

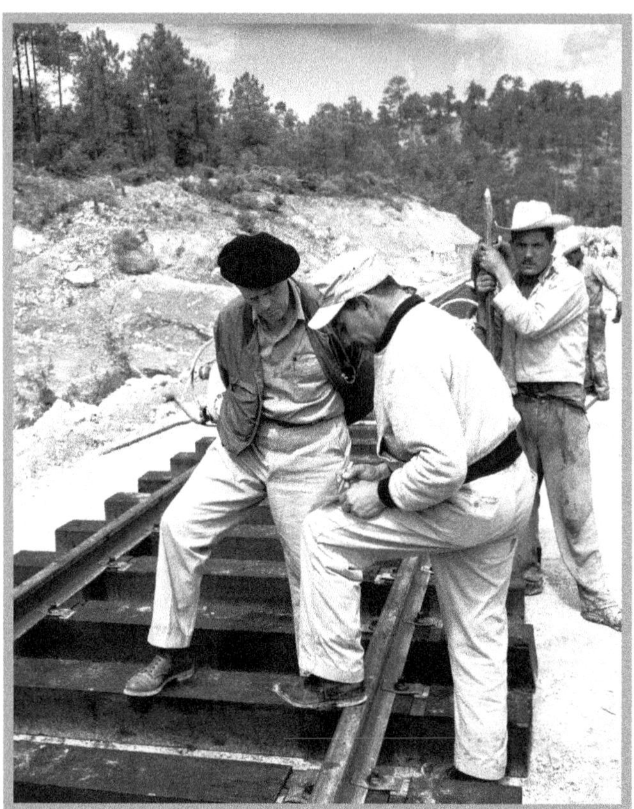

4.26. Chief of rehabilitation, Mariano García Malo, talks with visiting French railroad engineer Michel Streit who helped the Mexican engineers with the connecting of rails and the problems of expansion. Rails were beveled at the ends and overlapped (photo G.B. 1960).

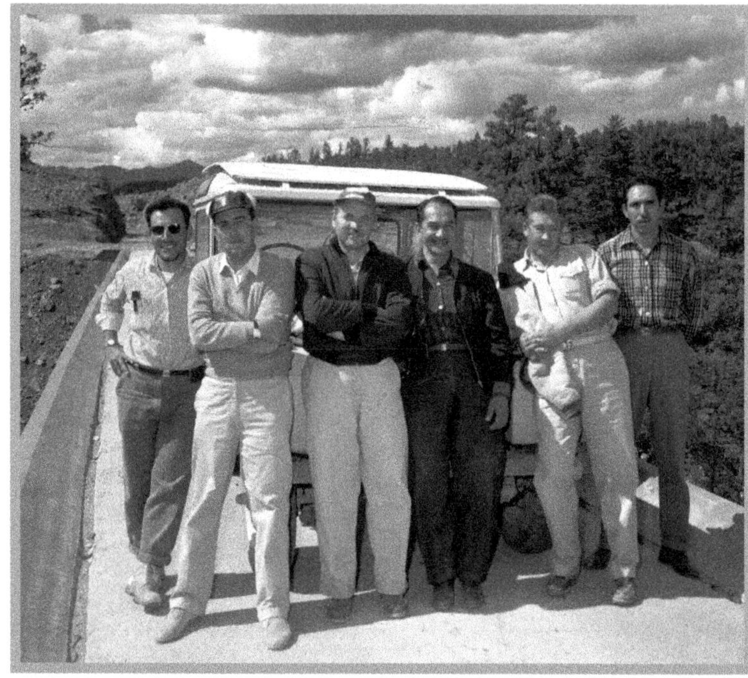

4.27. Engineers on a bridge. On the right is Ing. Duarte. Third from the right is Mariano García Malo, head of rehabilitation. The fourth from the right is perhaps Fernando Anzaldua (photo G.B. 1960).

FORT WORTH STAR-TELEGRAM

– Friday Morning, May 22, 1959

New Railroad in Mexico to Penetrate Sierra, Do 'Impossible'

by Glenn Burgess

CHIHUAHUA CITY, May 21— A.E. Stilwell, promoter of the Kansas City, Mexico & Orient, announced at a dinner in Kansas City in 1900:

"Gentlemen, the next railroad to be built in North America will be from Kansas City to the Mexican fishing village of Topolobampo on the Gulf of California."

But Stilwell had not seen the rugged Sierra Madre country he proposed to traverse.

So difficult for modern railroad building is the terrain of western Chihuahua and northeastern Sinaloa that Mexican civil engineers have had to add three years to their estimate of construction time.

One section of the region is so abrupt in its descent to the sea that Stilwell later announced a cog road with a 14 per cent grade and 40 miles long would have to be built there. Before he lost the project, however, he said he had discovered a 2.5 per cent route.

This route had to be rediscovered. Mexican engineers now have it—a cut-fill-tunnel-bridge route through incredible canyons.

Their achievement may go down in history as the top railroad engineering feat of North America.

Building 426 miles of the old Kansas City, Mexico & Orient, which was begun in 1900 and now is in operation as the Chihuahua al Pacífico to Creel, 189 miles west of Chihuahua City, was hard enough, but crossing the remaining gap of 180 mountainous miles became an engineer's nightmare.

The Sierra Madre at this point was originally a volcanic table land 8,000 feet above the sea. Weathering has carved the region into many deep canyons. American engineers tried to stay on top of the mountain ridges until they reached the Sinaloa State line, but at the end they always encountered the 2,000-foot drop that occurs in an extremely short distance west of the Chihuahua-Sinaloa boundary.

Pressure of World War II and Mexico's realization that rail connections between Sinaloa

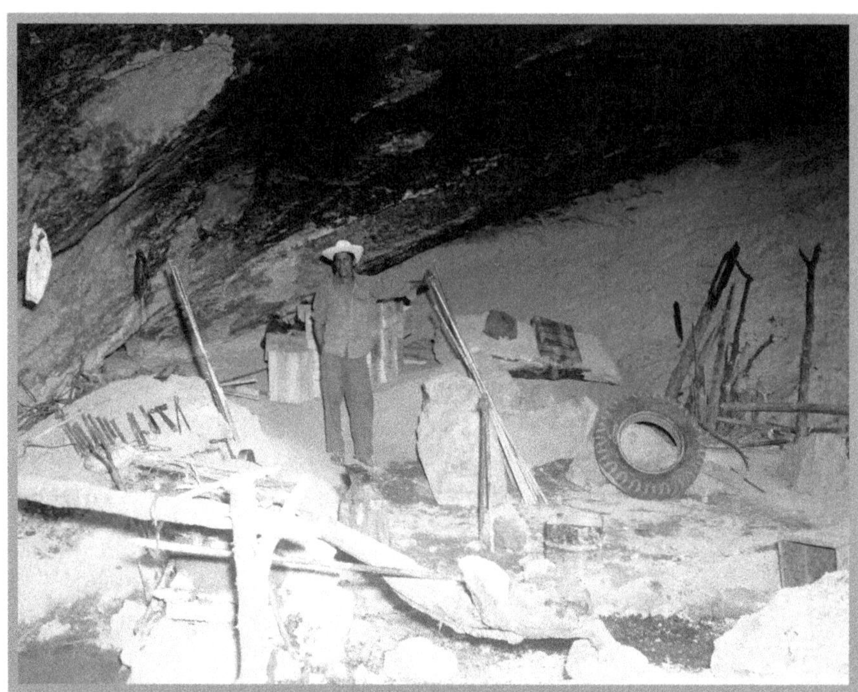

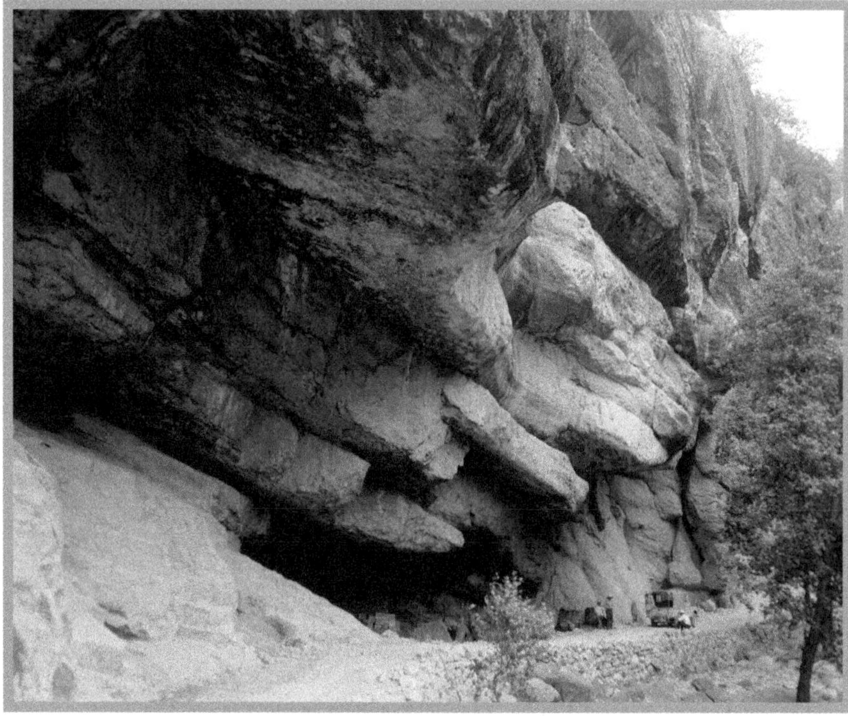

4.28 and **4.29.** The Cueva del Dragón (Cave of the Dragon), upstream from Cuiteco, was used for storing dynamite and for sharpening chisels with a charcoal forge (photos G.B. 1955).

and Chihuahua were vital to her future economy, caused Mexico to send her own engineers into the mountain wilderness. After years of work on foot and on horse and mule back, and without benefit of aerial or contour maps, Francisco Togno and his crews came out with plans that were acceptable to the Secretary of Communications and Public Works and to Mexico's president.

If the visitor who is privileged to go over the route now by pick-up, jeep, or truck, thinks the going is rough, he should remind himself that Togno used to make the trip every two months by horseback to visit his completely isolated engineering camps. The round-trip took him 12 days between Creel and Choix on the Rio Fuerte. Last month, the writer made this same round-trip distance of 302 jeep miles in four days and had time to visit many of the more interesting construction challenges.

Beginning at Creel, the grade takes off west from the Continental Divide and remains on top of the mountain ridges for 48 miles. This section has been completed in three years. Numerous tunnels, a cut 100 feet deep, a complete loop around a mountain with the railroad bridging itself, and a view into the 6,000 foot deep Barranca del Cobre, always will amaze the visitor.

Past the barranca and west of the village of Areponápuchi, the railroad begins its descent to the sea. First it drops into Cuiteco Canyon where 22 miles of grades are complete, as are most of the tunnels. From the Barranca del Cobre to Cuiteco, by rail, the distance will be 31 miles and the grade drops 1,712 feet from 7,247 to 5,525.

At Cuiteco, located in a circular valley, similar to the Chisos Basin in the Big Bend National Park, is one of the two division engineer camps with 40 civil engineers employed by the Secretary of Public Works. The work is in the charge of Fernando Anzaldua. Several of the "camp" houses are of permanent construction and will be section houses later. They have electricity and running water and are in daily communication by radio with Chihuahua City.

A half mile east of Cuiteco is the main construction camp of Ingenieros Civiles Asociados, S. A., contractors, who have 800 workmen in the Cuiteco Division. Many of the workmen have their families with them. The camp has its cafe, commis-

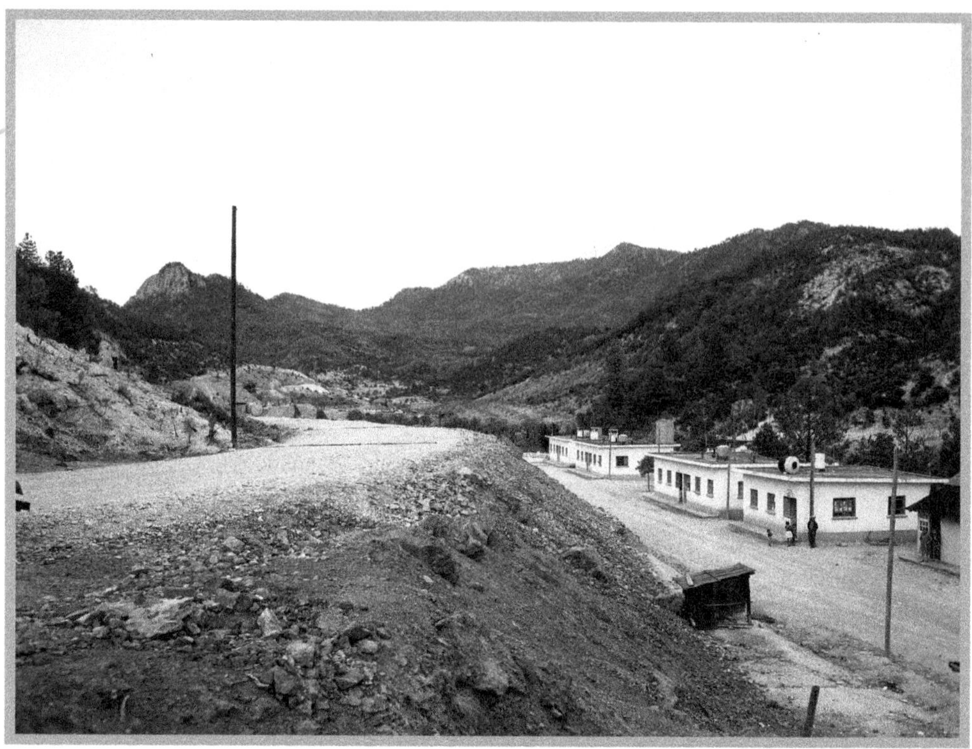

4.30. Railroad bed, Cuiteco office and homes for the SCOP engineers (photo G.B. 1955). The homes are still being lived in today.

Life at the Cuiteco Engineers' Camp

From my article in the *Alpine Avalanche*:

> We were fortunate in having almost modern houses with such conveniences as running water, toilets, etc. There is a central dining room where we would eat beans and tortillas for breakfast, dinner and supper; however, the food was fairly good as long as I stayed away from the chile.... [Ing. Anzaldua even talked a baker into setting up a business so we could have bread and pastries, but it did not last long.] Every Sunday at Cuiteco we would go to one of the nearby towns and spend the day playing baseball and basketball. On the last Sunday I was there, we all piled into the back of a truck and drove the four hours to the town of Creel. There we spent all morning playing baseball, which we won 26-4; rested an hour and played basketball, which we lost by 1 point in an overtime.

I was on a basketball scholarship at Texas Western College in El Paso at the time. A few years later, I played in a tournament with the Creel basketball team, in Creel, which we won. I was thrilled when asked to carry the Mexican flag in the opening ceremony.

sary, school, nurses, doctors, and basketball and baseball play grounds.

Tarahumara Indians working on the job simply use the narrow access roads for a ball game resembling the old game of "shinny."[1]

Cuiteco people grow apples. Before the coming of railroad construction it was three days by horseback from any road. Its church, of adobe, is one of the old Catholic missionary churches of the region.[2]

Cuiteco also boasts the first imposing and modern designed concrete bridge west of Creel. It is 87 feet high and 626 feet long. Below Cuiteco, the canyon is more of an open mountain valley. However, it still has its share of difficult tunnels, cuts and bridges.

The bridge, known as Cuiteco No. 2, is unique in all North America. The main beam, 87 feet above the stream, is of pre-stressed hollow concrete. Cables passing through tubes have a tension of 900 tons. When no train is on the bridge, the beam curves upward eight inches, but straightens out when trains cross. Of Belgian design, it is being studied by the Association of American Railway Engineers.

While the Cuiteco Division includes 85 miles of grades, actually only 31 miles are considered to be under construction.[3]

The adjoining 30-mile segment in the Témoris Division, known as the Rio Plata and the Septentrión canyons, still challenges the best in Mexico's engineers and construction men. It goes through what is undoubtedly the most majestic canyon of all Mexico, the Septentrión. This is the canyon that is delaying for three years more Mexico's dream of a completed Chihuahua al Pacífico railroad.

1 For a description of this game, a type of field hockey, see "Hit and Run" in *Natural History* and *Re'igí ra'chuela. (El juego del palillo),* both by Albino Mares Trías and Don Burgess McGuire.

2 The mission at Cuiteco was founded by the priest Juan María de Salvatierra in 1684.

3 Glenn's article "Tortuous Terrain Shattered Early Completion Hopes of Railroad" in the *El Paso Times,* May 15, 1959, which was only slightly different from the present article, added: "The construction completion chart, from east to west is:
First five miles—80 per cent complete.
Next 22 miles—100 per cent complete.
Last four miles—50 per cent complete."

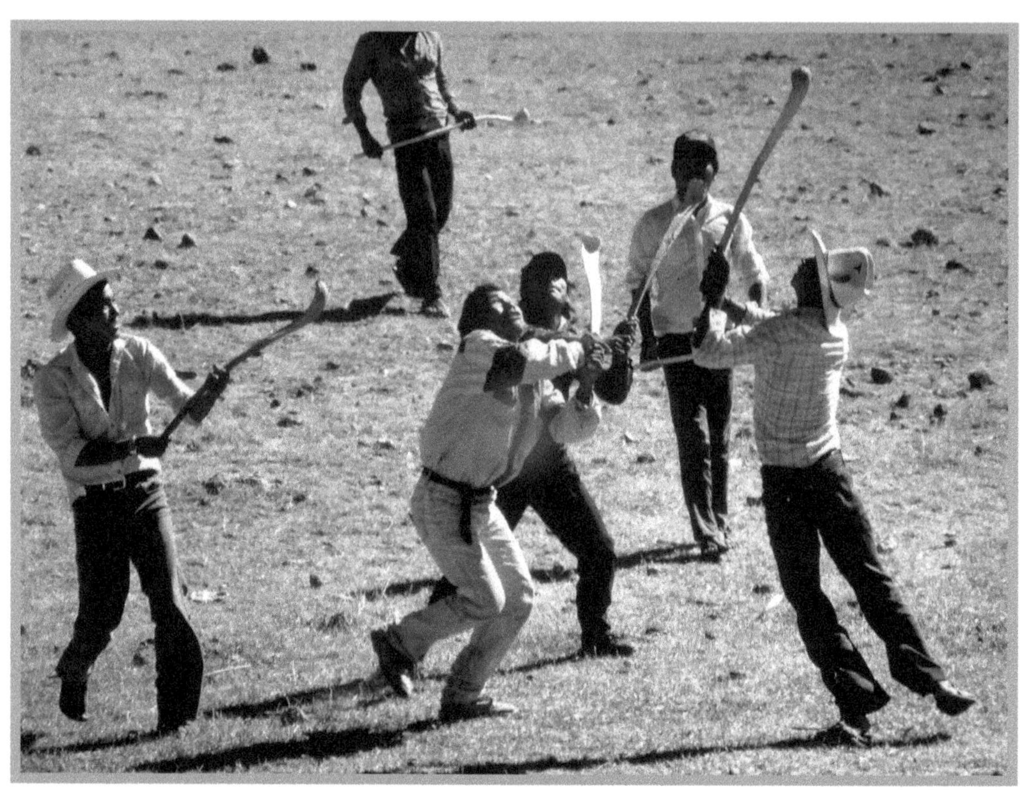

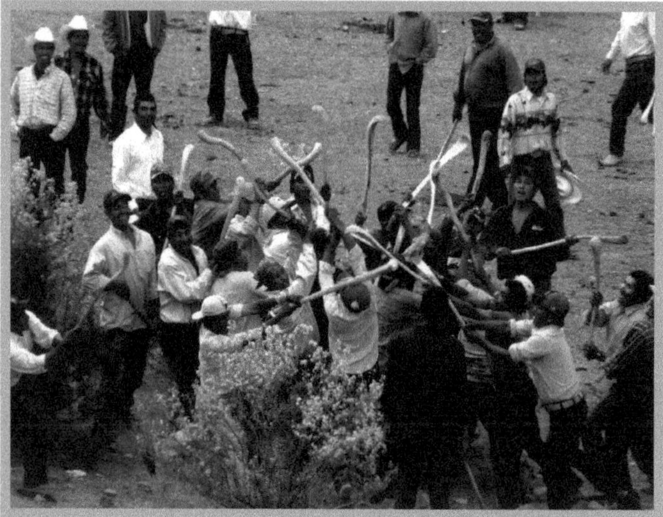

4.31 and **4.32.** Tarahumaras playing a type of field hockey called "ra'chuela" in Tarahumara and "palillo" in Spanish. The game is dangerous and hats are worn for protection. In the lower photo, note the golf-ball sized ball in the lower middle part of the photo next to a man's cap. (Photo above by D.B. at Rodeo ranch c. 1990; Photo to left by Andy Kjos at Cuiteco in 2008).

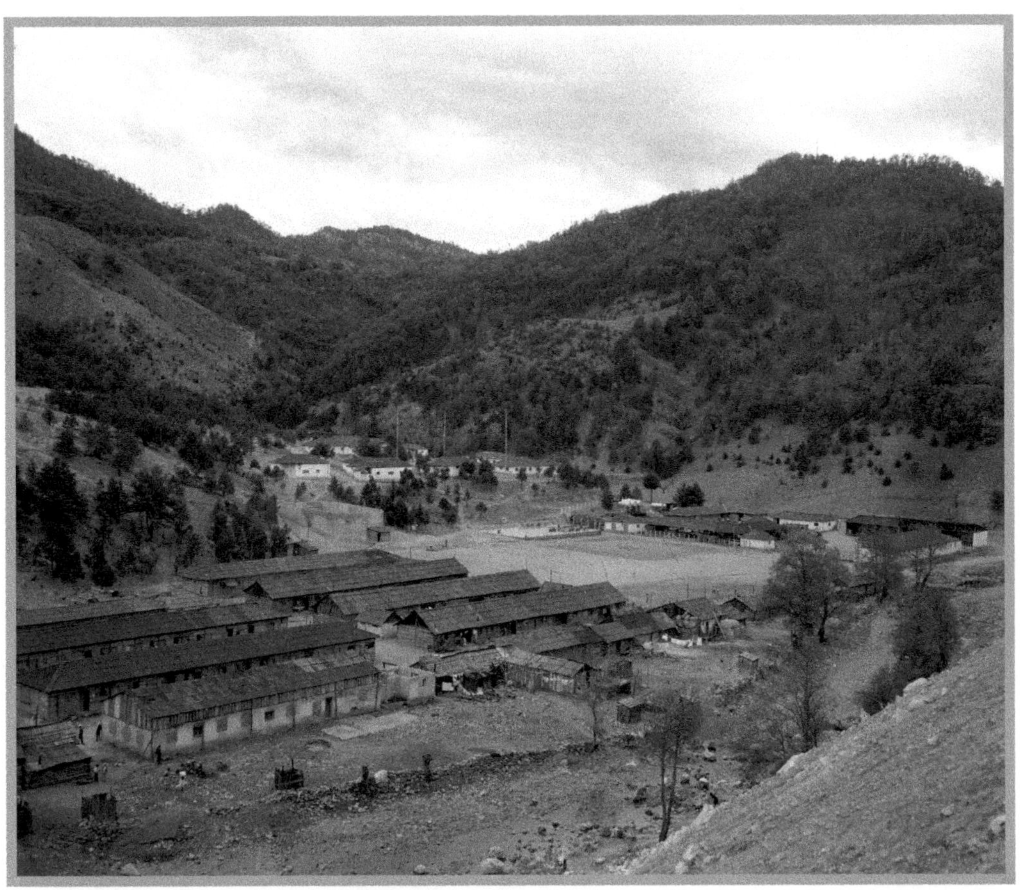

4.33. La ICA (*Ingenieros Civiles Asociados*) construction camp near Cuiteco called Oribo. White houses on the hill were engineers' homes. The building with no roof was the movie theater. The long houses were workers' quarters with the mess hall in the lower left. People remember a Chinese cook who ran the kitchen. Beyond these houses were the baseball field and the shops (photo G.B. 1955).

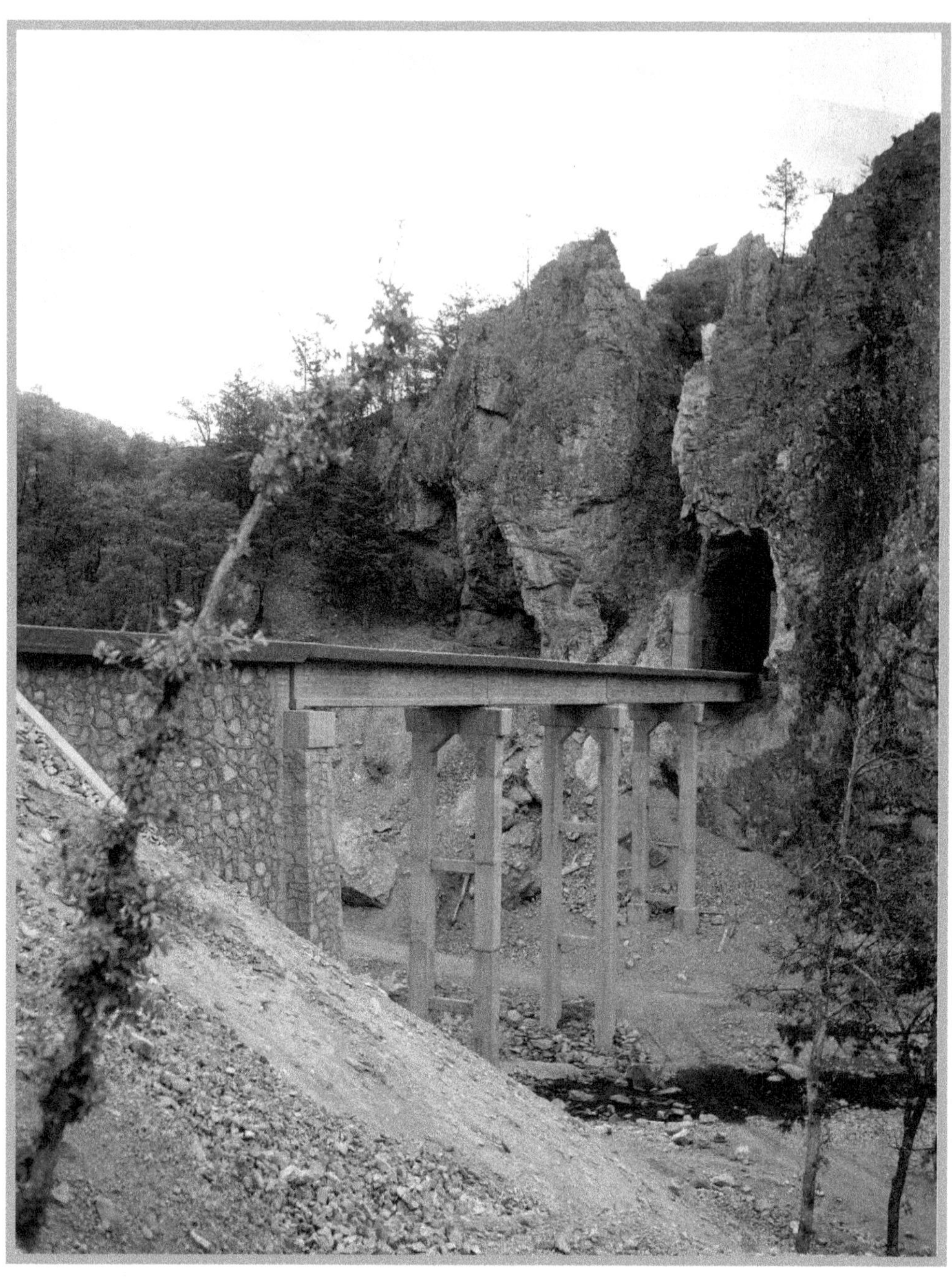

4.34. The bridge just downstream from the village of Cuiteco goes directly into a tunnel (photo G.B. 1955).

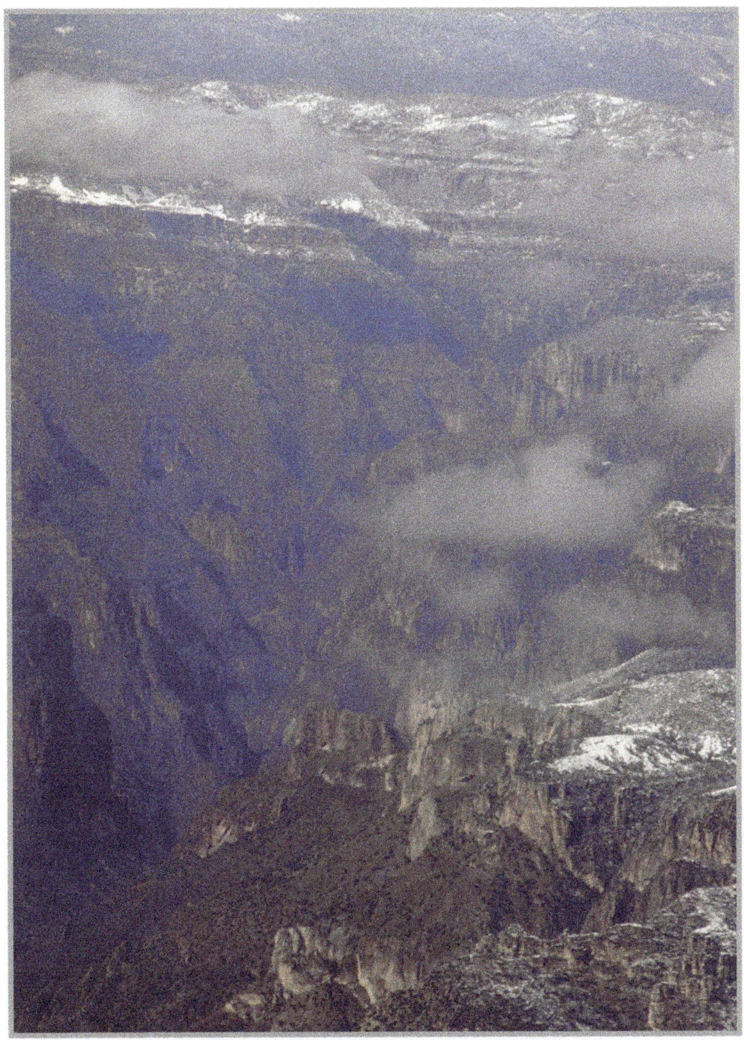

4.35. Snow in the barranca (photo D.B. c.1980).

The Barranca del Cobre

The Barranca del Cobre is part of the Urique Canyon. The name formerly applied only to a small section of the canyon where the Jesuits obtained copper for making their church bells.

As tourism has grown, the name has been used to apply to more and more of the canyon and the area in general. The word Urique comes from a Tarahumara word which means "hot country." The Tarahumaras themselves do not give names to canyons.

The depth of the canyon at El Divisadero is 4,265 feet (453 meters). Further down the canyon, six miles (10 kilometers) below the town of Urique, the depth is 6,135 feet (1600 meters). (These figures are from the geographer Dr. Robert Schmidt, Professor Emeritus at UTEP, *Mexico's Sierra Madre Occidental: The Geography*, forthcoming.)

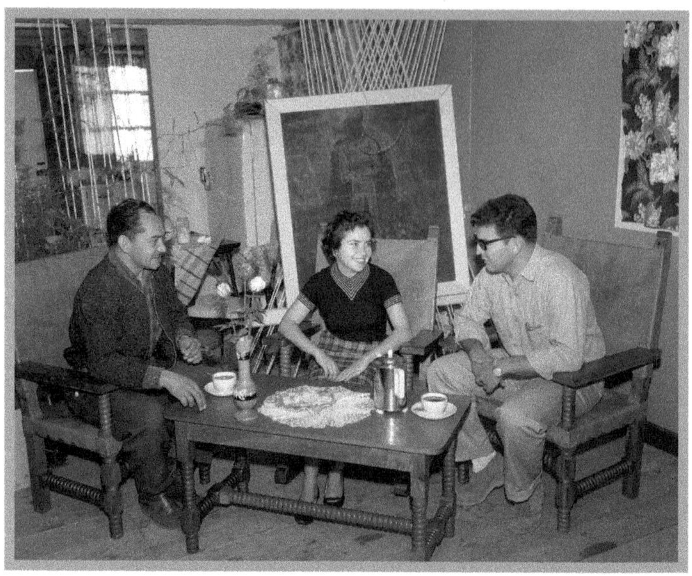

4.36. Mariano García Malo talking with Ing. Ernesto Talamantes, who directed the western division work, and his wife (photo G.B. 1960).

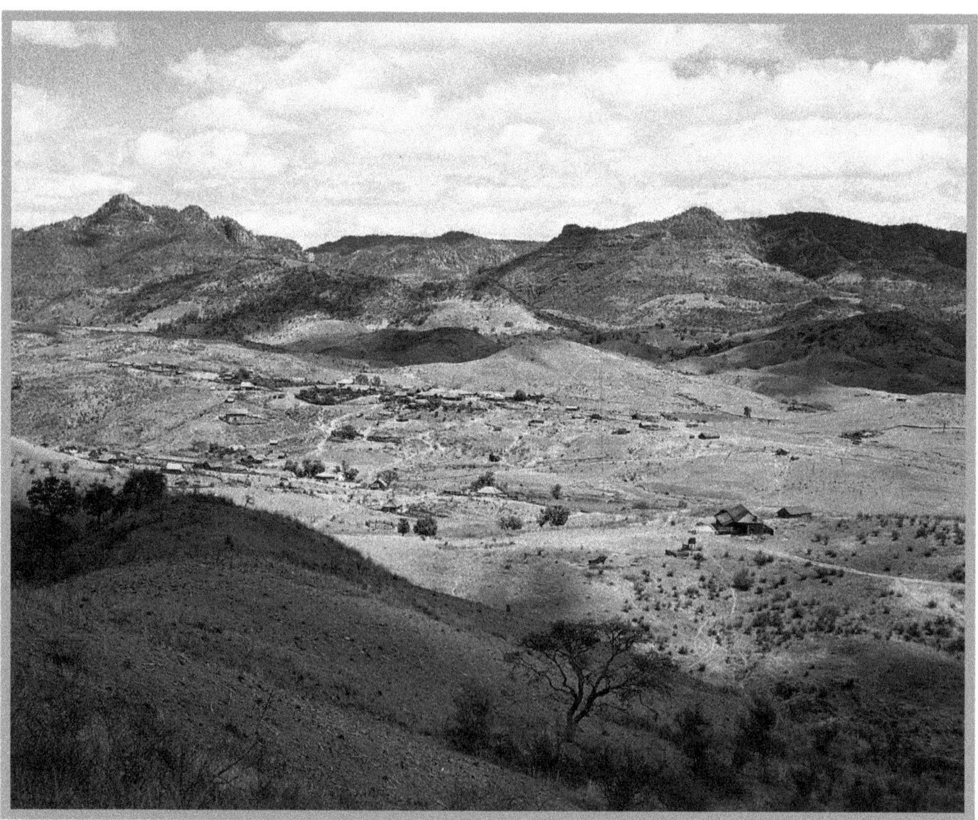

4.37. The village of Témoris, located above the canyon where Chicural is located (photo G.B. 1955).

The El Paso Times

- Monday, July 6, 1959

Final 30 Miles of Chihuahua Al Pacífico Railroad The Toughest

Engineers Defied Since 1900

by Glenn Burgess
Times Correspondent

Témoris[1], Chihuahua, Mexico—Only 30 miles of construction work stands between the first rough draft on the drawing board and the completion of the shortest rail route by 400 miles between mid-continent United States and the Pacific Ocean. This final 30 miles is already 40 per cent complete. The work yet to be done represents just 5 per cent of the grades, drainage structures, tunnels and bridges on the 180-mile section still without rails of the old Kansas City, Mexico & Orient Railroad.

It had defied engineers and construction men since 1900.

Percentage wise, it is an infinitesimal part of the 1,695-miles-long line between Kansas City and Topolobampo on the Gulf of California, however, the obstacles to be overcome are staggering.

Mexican engineers and two construction companies are gradually closing the gaps with completion dates on the road bed estimated at one to three years, depending on the amount of funds the president of Mexico can make available.

In the Western or Témoris Division, where A.E. Stilwell thought a cog road would be necessary over 40 miles of 14 per cent grade, the Mexican engineers have chosen to go down the Septentrión Canyon and at times, by doubling back in the majestic canyon, have been able to maintain the required 2.5 per cent grade. In this gorge, 285 miles southwest of

1 The actual town of Témoris is located in a valley on top of the canyon and the construction camp was located near the edge of the canyon, while the railroad runs in the canyon below. The train station is referred to as Témoris or Santa Bárbara, the name of a nearby mine. The construction project there was referred to as Chicural, the name of a local ranch. The name Témoris comes from the Tarahumara word for frog "remó".

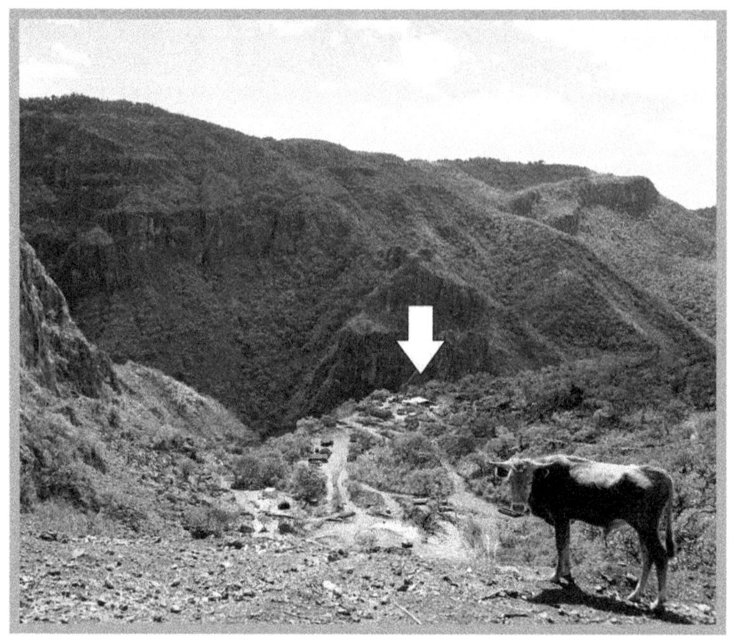

4.38. In the middle distance, the construction camp of the Águila Co. located on the edge of the Septentrión Canyon above the Témoris station (photo G.B. 1955).

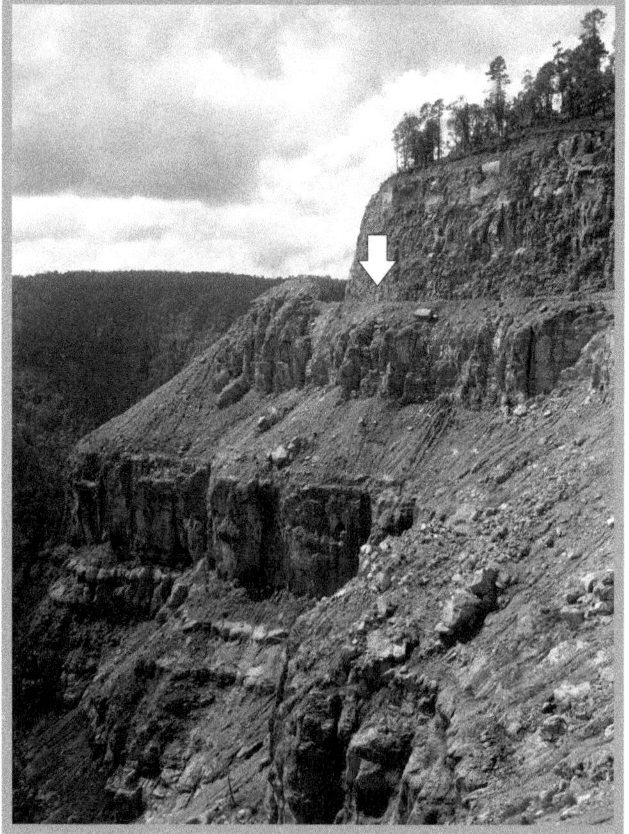

4.39. Shelf near Guasachique blasted out of a cliff. Ing. Leal tells how that once, after the line was in operation, there was a landslide at this point. The train engineer could not stop the train and it ran into the rocks. The engine was derailed and was hanging precariously over the edge. Ing. Leal and others were able to build a platform under the engine with railroad ties to keep it from going over the edge so that it could be pulled back (photo G.B. 1955).

Chihuahua City and near the Sinaloa State line, one section less than three miles long will have 10 tunnels and five bridges.

Another section, 1.6 miles long, will have 4.4 miles of track. Here the railroad, dropping down stream on the north bank of Rio Plata, goes first through a 50-yard tunnel, under a 100-foot waterfall, through a 50-yard tunnel, clings to the canyon wall for 400 yards, then makes an almost complete circle within a 3,890 foot tunnel called La Pera[2].

ANOTHER TUNNEL

Now it goes upstream several hundred yards to the waterfall, crosses the Rio Plata, turns downstream again on the south bank, goes through another short tunnel, crosses the river on a high bridge, and, before it disappears around another bend in the canyon, enters yet another long tunnel.

Locally, this engineering feat is known as the Chicural. It is about three miles from Témoris, headquarters of the Western Division work under the young Ing. Ernesto Talamantes. From Témoris a mining village in Western Tarahumara Indian Territory, the SCOP engineers direct the construction work along the Septentrión Canyon. Farther upstream it is known as the Cuiteco and Plata Canyons and the construction work is under the Cuiteco Division engineer.

Living at an elevation of more than 5,000 feet, the engineers drop down to 2,000 feet elevation in semi-tropical climate to plan and supervise the construction projects. One route goes southwest over a mountain ridge, then descends over 3,000 feet via jeep road to the river.

Doing the construction work is the Águila Co., employing 1,000 men. This year their contracts call for an expenditure of $3.2 million, the same as the Cuiteco work by Ingenieros Civiles Asociados, S.A.[3] Construction camps occupy the almost non-existing level spots along the bottom of the canyon. In some locations, houses for the workmen will have to be removed before the grade can be completed.

The Águila Co. has modern machinery and is well experienced in this type of terrain. It began six years ago at the end of the rails in Sinaloa and has gradually penetrated the rug-

2 Schmidt's latest map gives this length as 3074 ft.

3 Locally called La ICA.

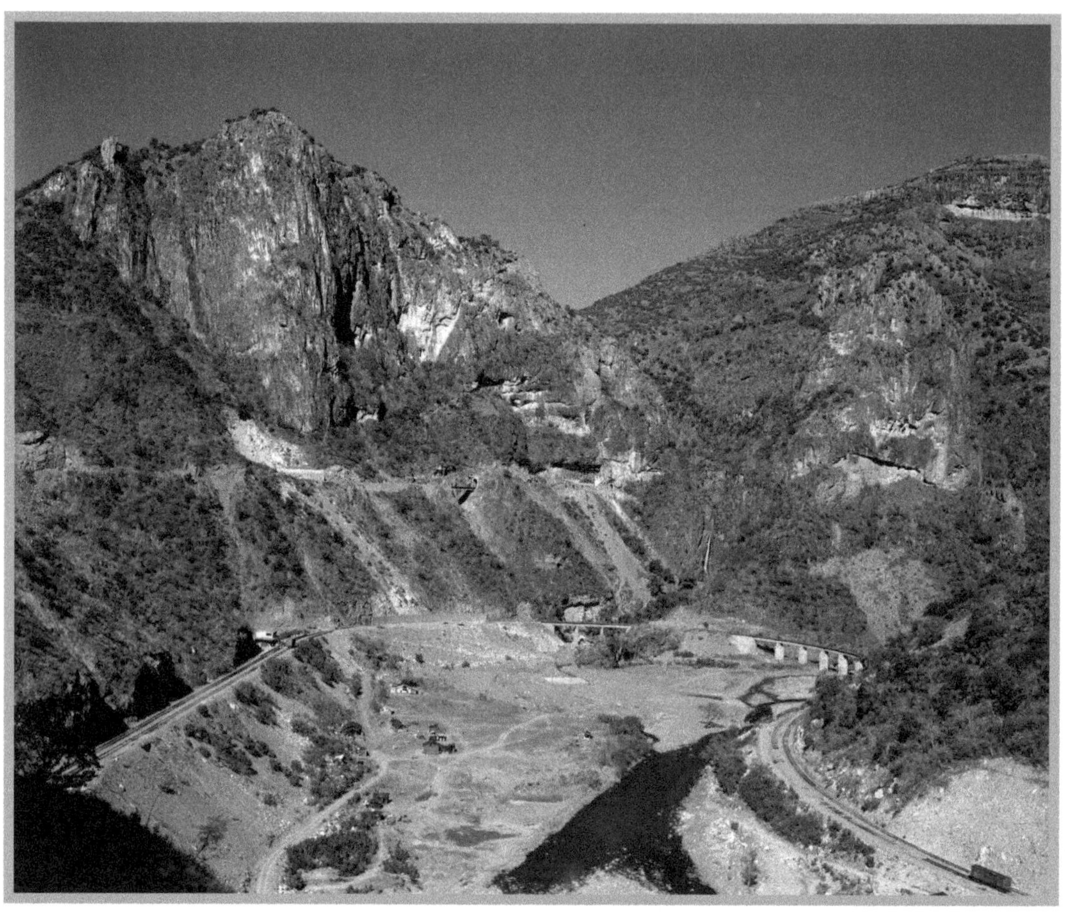

4.40. Chicural: there are 4.4 miles of track in a linear distance of 1.6 miles. View is upstream in Septentrión Canyon (photo G.B. 1960).

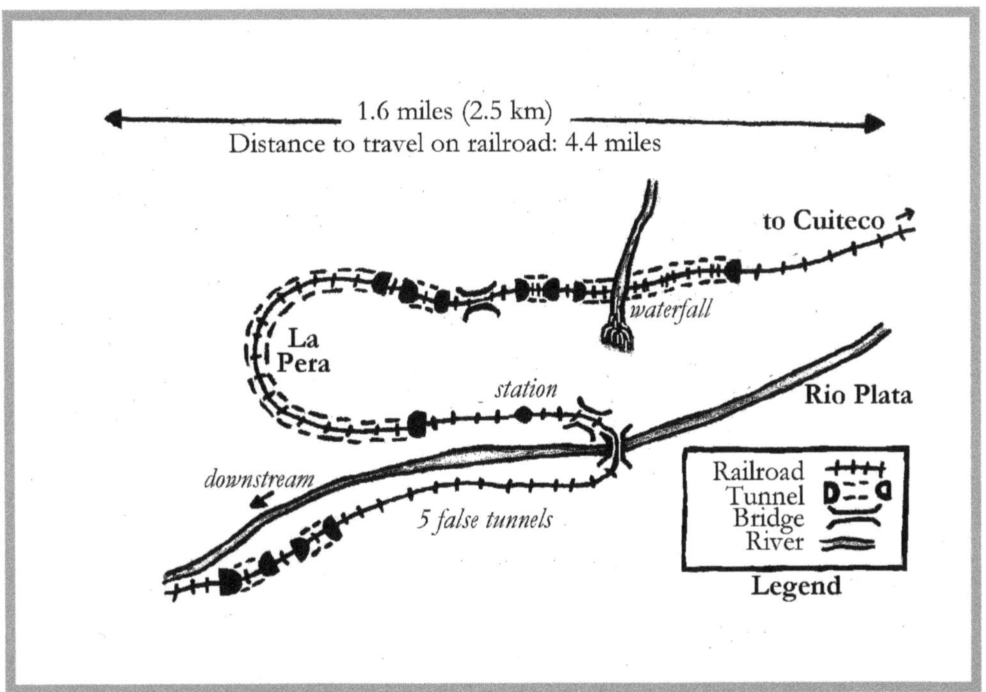

4.41. Sketch of Chicural

Chicural

This engineering feat consists of a close series of tunnels and bridges. According to Ing. Yáñez, starting at the upper end, there is a tunnel 818 meters long which goes under a waterfall, then a tunnel 40.3 meters long, then a short bridge, then a tunnel 189 meters long, then La Pera Tunnel which he lists as 946 meters, then the Témoris Station, then a bridge about 265 feet long, then the 714 foot bridge over the Rio Plata which was made in a curve. And then there are five false tunnels whose lengths are 126.29 meters, 299.6 meters, 316.5 meters, 200.3 meters, 122.8 meters, followed by two regular tunnels of 96.2 meters and 84.3 meters long. It needs to be noted that there are sometimes differences in the numeration of the tunnels and on the lengths of the tunnels and bridges. Part of the numeration problem is caused by the false tunnels that have been added since the original construction. The different lengths of tunnels and bridges could also be caused by different approaches in deciding exactly where a tunnel or bridge begins and ends.

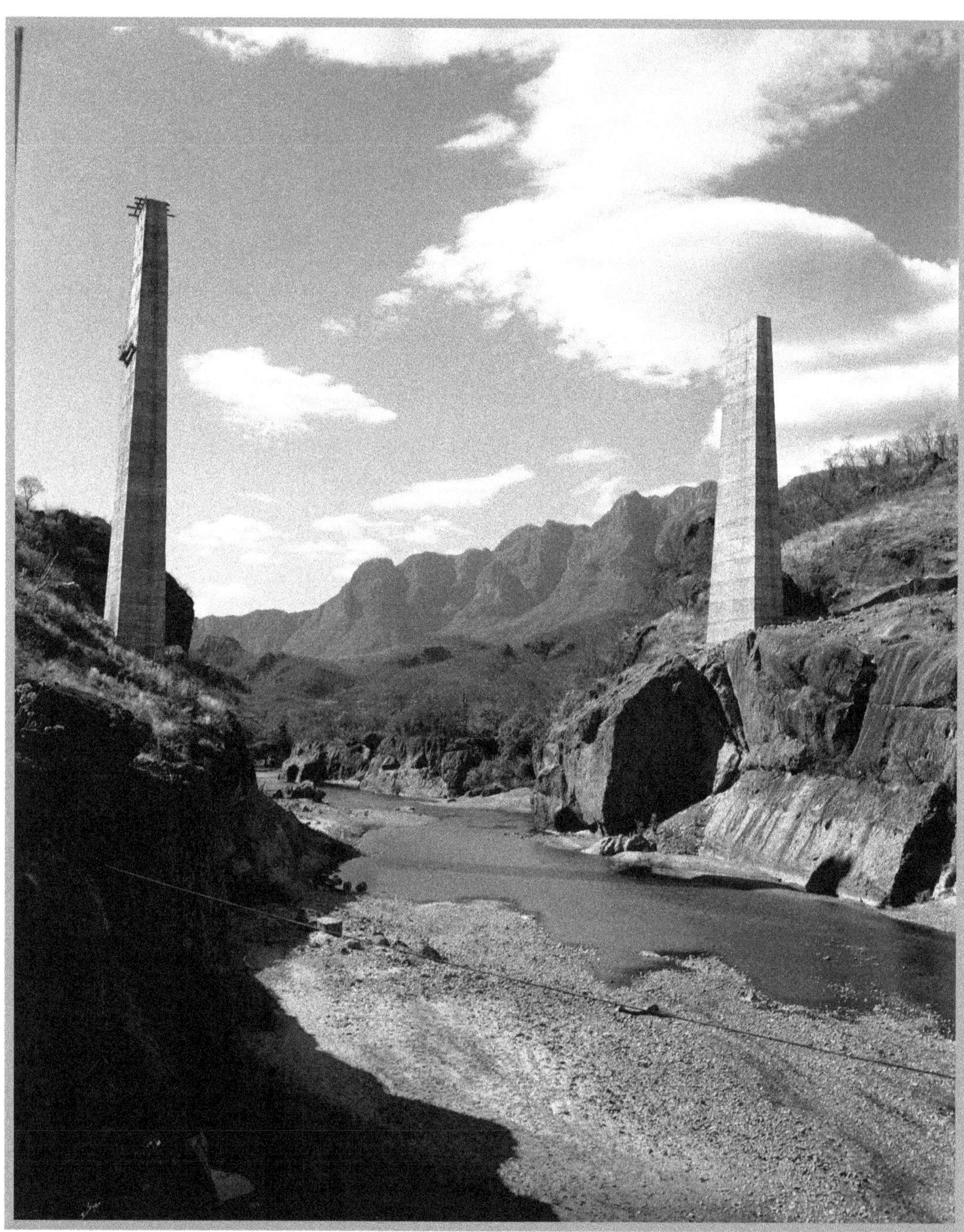

4.42. **RIDE HIGH**--The tracks here will be 300 feet above the Rio Chínipas. In the background is a 2,000 foot drop that confronted early engineers at the Sinaloa state line. Finally, they chose Septentrión Canyon as the route out of the mountains. The canyon ends about two miles from the bridge (photo G.B. 1955).

ged mountains. The grades alone have cost $34,480 per mile.

In the 25 miles around the Chicural section, 15 tunnels will be required. These must be driven through very hard volcanic lava flows and intrusions—costly work in any man's country. It is dangerous work, too. In the upper Chicural section, first an access road from Témoris drops down into the canyon to the old Santa Barbara Mine. Then, at least a mile-long shelf had to be blasted out of almost perpendicular canyon walls.

Next begins the painful drilling and blasting to make a solid rock shelf for the railroad. Workmen, like ants, creep along the face of the cliff, drilling holes with their compressed air drills to obtain foot-holds. These are expanded until machinery can operate and the compressed air machines and air lines can be moved a little farther along.

The work goes steadily on, and as far as construction projects go, few men have lost their lives.

Longest tunnel on the railroad is seven miles east of Rio Chínipas Bridge in Sinaloa. It is 1.2 miles long[4] with vehicle traffic moving through it without benefit of a signal system. A pinpoint of light can be seen at the opposite end. Drivers do not enter if they see a headlight. It is impossible to pass in the tunnel. If tunnels are costly, so are some of the bridges. Take the one across the Rio Chínipas, one mile west of the Sinaloa-Chihuahua state lines. The two concrete-steel towers are 211 feet above the solid rock foundations. When the steel structures are completed this fall, the span will be 292 feet long, and the rails will be 300 feet above the river—as high as the Southern Pacific Bridge across the Rio Pecos.

114 FEET HIGH

Twenty-two miles west of the Rio Chínipas is another bridge crossing the large Rio Fuerte. The concrete towers are 114 feet high and the bridge will be 1,454 feet long compared to the 880-foot long structure at Rio Chínipas.

Stabilizing the mountainsides after the railroad grade is completed is one of the major problems. Sometimes, stone and concrete walls keep the grade from slipping down the mountain. Again, after tunnels are built, hidden fractures may make it necessary to line the tunnel with concrete and steel.

4 The Ferromex office lists this bridge as 6,024 feet long or 1836 meters. Schmidt's latest map puts it at 5966 feet or 1820 meters.

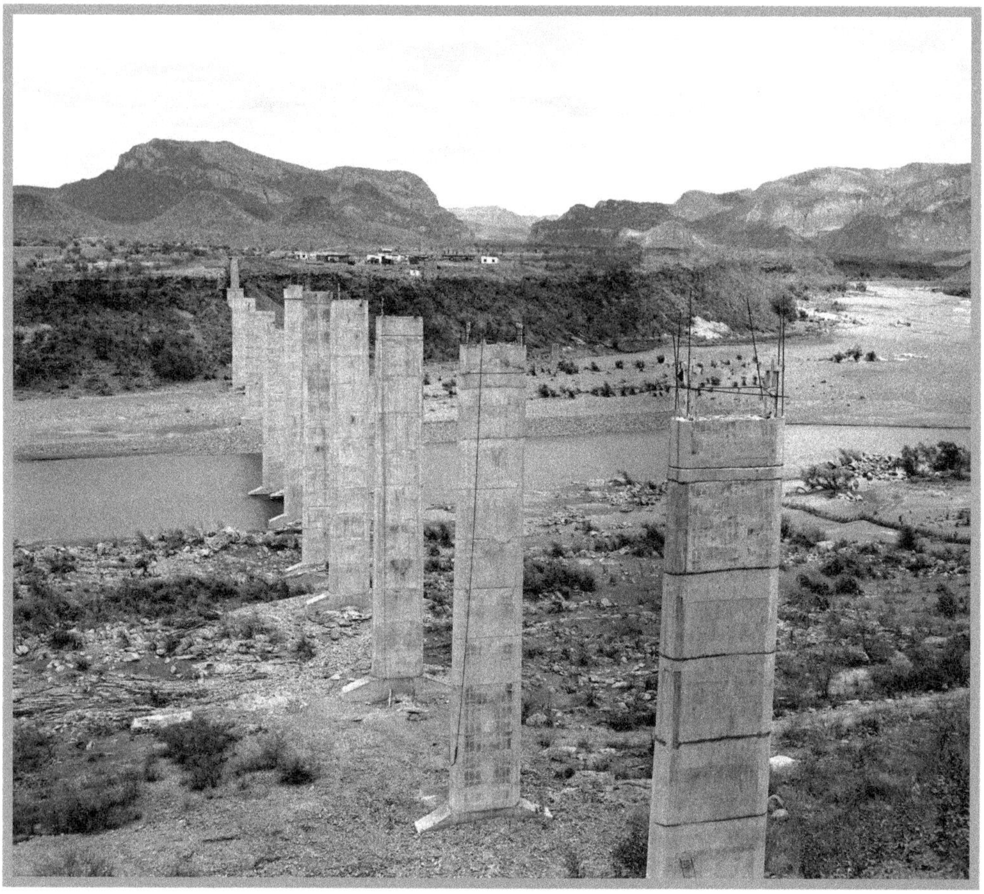

4.43 and **4.44.** Bridge construction over the Fuerte River. Ing. Leal tells how that when the top pieces of the bridge were being put on, he was up on top working and tripped. To keep from falling off the bridge, he grabbed a hot compressor air tube. This saved him from falling off the bridge but he suffered several burns. In the excitement, he said, he did not feel the burn until afterwards (photos G.B. 1955).

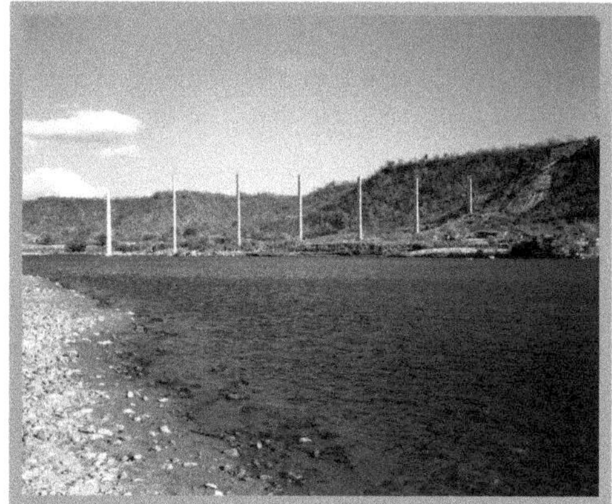

Rock slides can be devastating and are a menace for several months after the cuts are made.[5]

Five miles east of the Rio Chínipas Bridge is a single rock slide. Last September, when the unprecedented rains came, one lone rock rolled over on a shallow cut. It is shaped like an egg, is 75 feet high and weighs just 9,000 tons.

During the rainy season, heavy rains fall on the western slopes of the Sierra Madre. A few miles below the Rio Fuerte Bridge, is one of the major reasons why the Chihuahua al Pacífico, the new name for the Orient, will develop heavy tonnage—the Miguel Hidalgo Dam. Completed three years ago, it stores 7,000,000 acre feet of water. Added to the regular flow of the Rio Fuerte system, it irrigates 500,000 acres of land in a rich coastal plain with a semi-tropical climate. When fully developed, 350,000 people will make their living off this dam.

Three huge generators will have a total out-put of 60,000 kilowatt hours of electricity. The first is now being installed by an American company. The armature came from Japan, purchased with Mexican cotton.

At peak flood last fall, the Rio Fuerte furnished 14,000 cubic meters per second. More water went over the spillway than the huge lake holds. This convinced the hydraulic engineers that the original design of the earthen dam should be completed soon by adding an additional 32 feet. Then the dam will be two miles long, 250 feet high, and will irrigate 700,000 acres. If this is not enough water for irrigation and power, a short distance above the upper end of the artificial lake is a narrow opening, Huites Canyon[6], where the Rio Fuerte, that collects its water from four states, emerges from the Sierra Madre.

Preliminary studies reveal a concrete dam 570 feet high can be built here.

Eight miles southwest of the Miguel Hidalgo Dam, across the fertile flood plains of Sinaloa and past San Blas and the rapidly growing agricultural town of Los Mochis, remains the Port of Topolobampo with its 11-mile long bay that averages 270 feet deep.

5 I know of at least one tunnel, near Cuiteco, that collapsed and was made into a cut.

6 The Jesuit priest Andrés Pérez de Rivas, writing in 1644, states that the Huites were an indigenous group "whose name signifies the use of bow and arrow, at which they are extremely dextrous" (translation of Thomas Robertson, p.77).

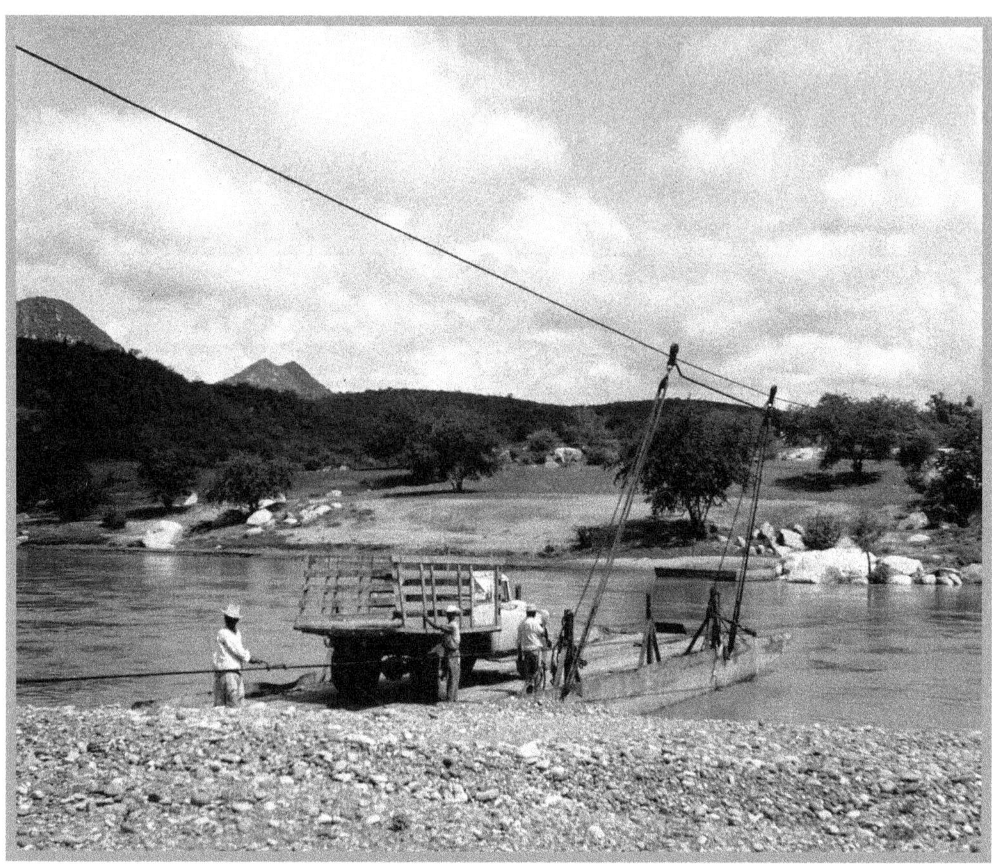

4.45. **HUGE VOLUME**—Even in dry weather, the Rio Fuerte furnishes a huge volume for the Miguel Hidalgo Project. During the flood last September it poured 14,000 cubic meters per second into the man-made lake. This project is tied in with the railroad construction (photo G.B. 1955). [These barges were known as "pangas" or "chalanes".]

GREAT CITY

Wrote Stilwell in 1912: "The port of Topolobampo will be one of the great cities of the Pacific; it will have its line of steamers to the Orient, Central and South America, New Zealand and Australia."[7]

Also wrote rather skeptically L.L. Waters, Faculty of the School of Business, in his *Steel Rails to Santa Fe*, in 1950: "From time to time, reports of the Mexican government intimate that a portion of Stilwell's dream will be realized. The Orient conceivably could be built over the Sierra Madre. The remainder of the promoter's dream, dealing with the heavy volume of freight, would still remain far from reality."

If Waters lives three years more, he may hear of trains running over the entire route. Local Mexican engineers working on the project feel it will be three years before they are ready to lay the rails.

Mexico could borrow the money to complete it sooner, but it prefers to take a little longer and use its own money. However, Ing. Francisco Togno, chief of construction for the Railroads of Mexico, who knows intimately every meter of the route through the Sierra Madre, said in Mexico in early June, that it is still possible that Mexico can provide enough funds to complete the grades, tunnels, and bridges by June 1, 1960.

Mexican engineers believe in the future of their railroad and declare that if Waters lives just another 10 years, he more than likely will see Stilwell's prediction of a heavy volume of national and international freight come true.

Old timers at Presidio are wondering, too, what will be the effect on their little town. They saw the original road go into the hands of receivers in 1912. In 1917, the receivers sold it to William T. Kemper of Kansas City, Clarence Histed and others for $3 million. Then, after the McCamey oil boom boosted its value, Kemper and his group sold the line to the Santa Fe in 1928 for $14 million, definitely

7 During the Stilwell time, the people in the USA started calling the Port of Topolobampo, Port Stilwell, and the Mexicans were visibly upset. Stilwell already had one port named after himself—Port Arthur, Texas. Wasn't that enough? Even the railroad was often referred to as the Stilwell Railroad (See *The Engineering and Mining Journal* (July 1907), and Almada, *El Ferrocarril de Chihuahua al Pacífico* (pp.126-127). Almada says that an article published in *El Norte* in Chihuahua City (Nov. 27, 1902: # 1069) stated: "It should be noted that in Mexico there is no place named Port Stilwell and the one that the American pirates want to call that, has been and will be, as long as our Government does not decree something different, the port of Topolobampo."

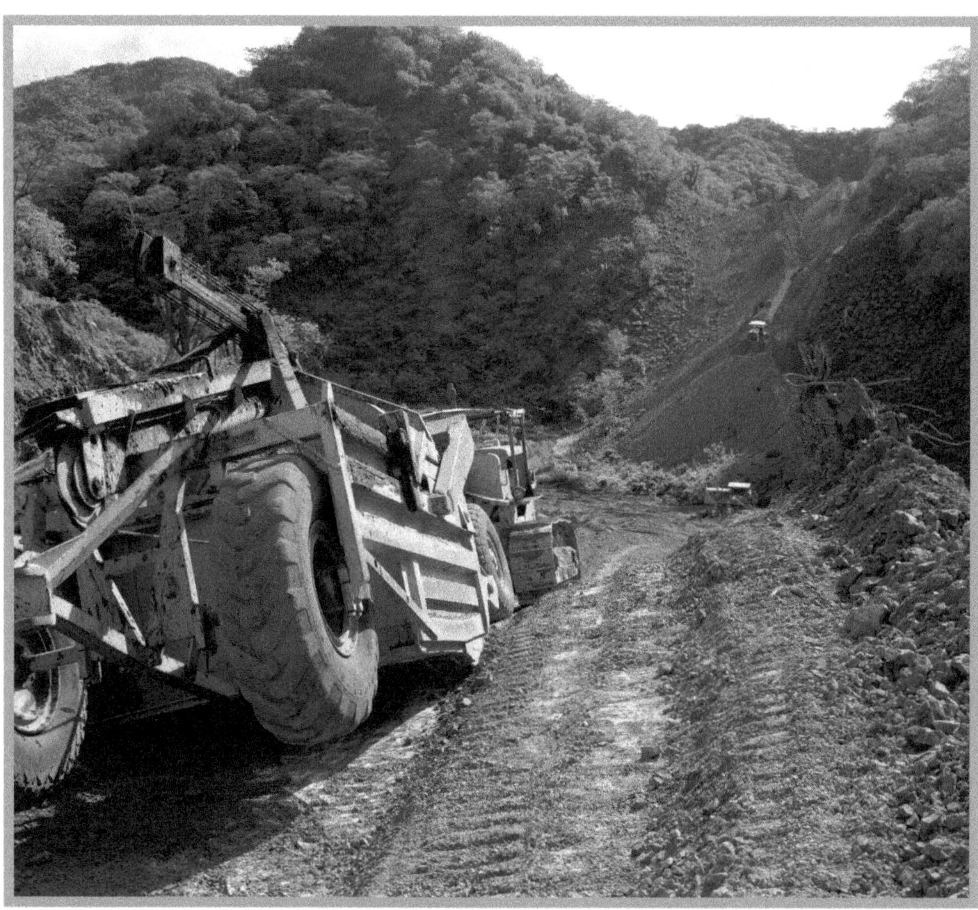

4.46. **HEAVY CONSTRUCTION**—Modern machinery of the Águila Construction Company, which has the contract to build the western division of the Chihuahua al Pacífico Railroad, slashes a tough trail (photo G.B. 1955).

a defensive move on the part of the Santa Fe.

SOLD BY SANTA FE

Santa Fe sold the 436 miles of railroad and 180 miles of unconstructed route through the mountains in 1929 to B.F. Johnston, a sugar operator for $650,000 cash and a mortgage of $900,000.

By Nov. 2, 1930, the Santa Fe had built from Alpine to Presidio and a special train ran to Chihuahua City. The people of Presidio thought prosperity had come. But the depression deflated the boom overnight and delayed the completion of the railroad for 30 years.

Hard times caused Johnston to have to default on his debt and in 1940 the Mexican government took over in conformance with its program of nationalizing the railroads. Later a token payment of $90,000 was made by the Mexican government to the Santa Fe in full discharge of the obligation.

Now observers predict the completion of the railroad and of a first class highway from Presidio to Chihuahua City about the same time will improve materially Presidio's and Ojinaga's economy.

For Mexico, the completed railroad is expected to remove isolation in a large area and develop the forest, mining and tourist resources of the Sierra Madre. It will provide an artery for freight and passenger traffic between two of its richest states. Any development of world trade and trade between Mexico and the United States will simply mean additional revenue to a country that has the vision and ability to complete the dream of A.E. Stilwell and others who began an almost impossible project in 1900.

The Fuerte River Valley

George Simmons in his report to Owen in 1881 (*The Texas, Topolobampo & Pacific Railroad and Telegraph Company: Reports*....p.10) stated that the Fuerte valley is "magnificent in beauty—unsurpassed in fertility—the garden of Mexico—a land so rich and bounteous that if the proper measures to secure a sufficient amount of it shall be taken, I am confident that the entire cost of the road, from Topolobampo to the foot-hills of the Sierras, can be defrayed from the profits which will accrue to the Company from future land sales."

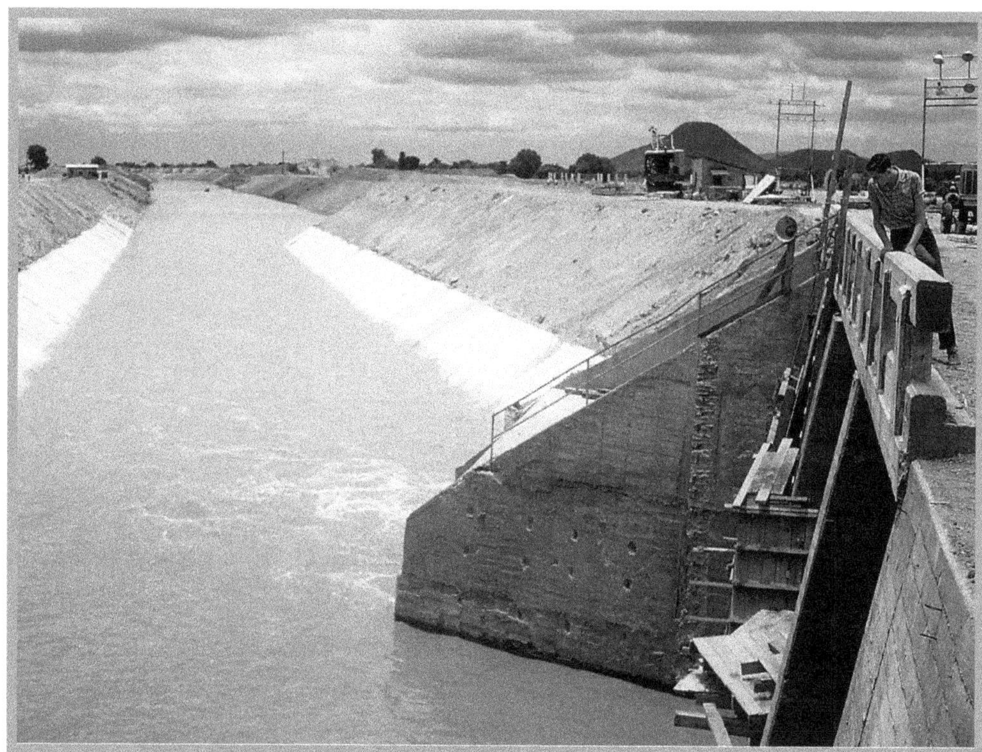

4.47. **WATER FOR 500,000 ACRES—**One of the two main canals from the Mahone Dam, which will irrigate a half-million acre agricultural project in Sinaloa on the route of the Chihuahua al Pacífico Railroad (photo G.B. 1955).

- Date unsure

Over Rugged Terrain
Mexican Rail Project to Tie Texas with Pacific

Editor's Note: The accompanying story, on construction of the railway from Ojinaga, opposite Presidio, Texas, to Topolobampo on the Gulf of California, was especially written for *The Star-Telegram* by Glenn Burgess of Alpine, who has just returned from a trip over the 587-mile route.

Incidentally, there is some doubt just how long the railroad will be. Mexican engineers admit that the sum of miles in the sections yet to be built, plus those already in operation, is 596 miles, but they think the discrepancy is somewhere in the new section mileages.

BY GLENN BURGESS

The largest natural harbor on the west coast of North America between the Panama Canal and San Francisco is Topolobampo Bay on the Gulf of California, which will be the western terminus of the Chihuahua al Pacífico Railroad.

The 587-mile line from Presidio, Texas, to tidewater in the Mexican state of Sinaloa, will bring to reality—perhaps within three years—the dream of Arthur Stilwell, promoter of the Kansas City, Mexico & Orient Railway.

The Orient line in the United States was acquired by the Santa Fe System and now is operated in Texas by the Panhandle & Santa Fe. Having taken over the Orient route across the border, the Mexican government is carrying it to completion through the Secretariat of Communications and Public Works, or the SCOP.

Most rugged of the 175.95 miles of construction required to complete the railway is on the western division, where the Águila Construction Company is working with modern equipment toward the northeast. The company is the largest of its kind in Mexico.

Grading and drainage work is 38 per cent completed on the eastern division, from Creel to the Septentrión Canyon, a distance of 111.38 miles.

NO JEEP ROAD

There is no jeep road through this narrow deep canyon that is 21.70 miles long, and the only communication between the eastern and western divisions is by plane or horseback. However, the Águila Company expects to have a jeep road finished through the canyon within a year.

The canyon has been surveyed and the cross section work completed. There will be a terrific drop from Témoris village into this beautiful gorge.

At its lower, or southwestern end is Las Guazas, at the junction of the Septentrión and Chínipas Rivers, approximately 133 miles southwest of Creel, railhead on the eastern division, and approximately 43 miles northeast of the western railroad town of San Pedro.

There is train service from Ojinaga, across the Rio Grande from Presidio, to Creel, a distance of 352 miles, and from San Pedro 68 miles to Topolobampo.

Beginning at Las Guazas and heading southwest there is a section 8.3 miles long under general construction, including a tunnel 5,966 feet in length [6,024]; a 10.54-mile section on which grading and drainage is 95 per cent complete and a 24-mile section on which the roadway is completed except for an expensive bridge across the big Rio Fuerte.

Águila's main construction camp, called La Mesa, is located on the Fuerte River, approximately 112 miles northeast of Topolobampo.

The company is pushing its giant earth-moving machinery forward toward the northeast at the rate of 1.55 miles per month.

HYDROELECTRIC DAM

A couple of miles from Camp La Mesa, at the mouth of the Huites Canyon, the Mexican government is planning to build a hydroelectric dam that will be 600 feet high. Águila hopes to bid on this contract when the time comes.

The company and the government agency, SCOP, expect to complete a huge irrigation dam within six months at a point on the Fuerte River eight miles northwest of San Pedro.

Up to 5,000 men have been engaged in moving 12,000,000 cubic meters of earth and stone to erect this dam, which will have a maximum height of 216 feet and a length of 1.86 miles. The project will cost 160,000,000 pesos.

This dam, called the Mahone or San Miguel, already is

backing up water and the reservoir it will create is expected to be filled in one year.

Government statisticians say the project will create a living for 310,000 persons.

Between the dam and the Gulf of California is an extremely level flood plain, with just enough fall to drain its deep, rich soil. The plain receives insufficient rainfall for cropping, however, and is covered with brush and cactus, which will have to be removed. Clearing the land and getting it ready for irrigation will cost an estimated $22.40 per acre.

The Fuerte is one of 12 rivers that come out of the high Sierra Madre and cross the state of Sinaloa. The only other stream that is dammed is the Tamazula River, in the Culiacán area, but the remaining 10 can be employed for irrigation and in time will be.

About 70,000 hectares, or 175,000 acres, in flood plain have been put into cultivation with irrigation by canals from the Fuerte, supporting a present population of 150,000 in the valley.

100 MILES OF CANALS

The Mahone or San Miguel Dam will bring under irrigation an additional 200,000 hectares, or 500,000 acres. To distribute the water there will be 100 miles of major canals and 124 miles of lateral canals.

It is anticipated that 70 per cent of the farming operation will be mechanized and that settlers will be brought in from all over Mexico, but it is believed Sinaloa will supply 80 per cent of the farmers.

The valley now produces cotton, sugar cane, wheat and tomatoes. The tomato crop matures earlier than any in the United States and commands a good price. Corn and fruit also are produced.

The climate is semi-tropical.

The town of Los Mochis, on the Chihuahua al Pacífico, 12 miles from Topolobampo, is expected to benefit most from the agricultural development. Established in 1902, it now has 30,000 inhabitants. Its wide streets and activities resemble those of earlier towns in the Texas Panhandle and on the South Plains.

The Chihuahua al Pacífico Railroad is due to develop an agricultural area from Ciudad Obregón, 150 miles north of Los Mochis, to Culiacán, capital of Sinaloa, 124 miles south.

VALLEY FERTILE

Arturo R. Murillo, general manager of the Bank of Culiacán, declares the valleys

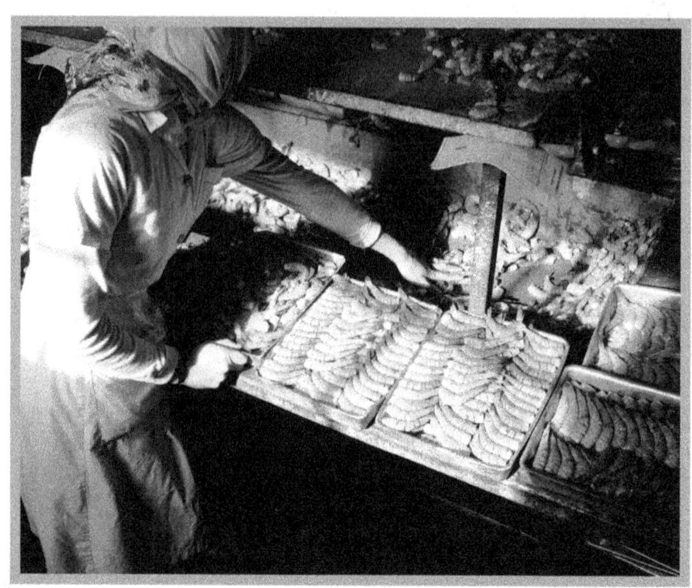

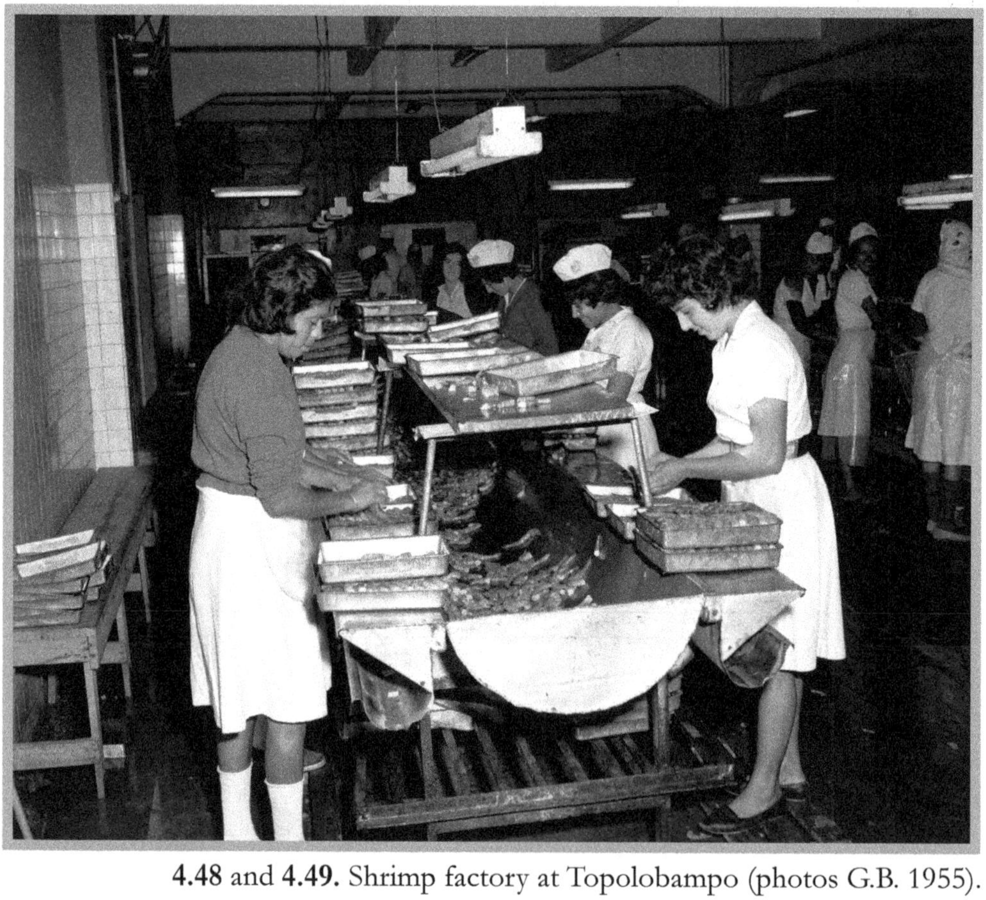

4.48 and **4.49.** Shrimp factory at Topolobampo (photos G.B. 1955).

around Culiacán, irrigated from the Sanalona Dam on the Tamazula River, produce crops worth 140,500,000 pesos and the Fuerte Valley output is valued at 145,027,000.

Murillo mentioned that the state of Sinaloa is rich also in the production of lead, zinc, copper, silver and gold.

Topolobampo, whose name of Mayan[1] origin means "a turtle (presumably of a giant sea species) was seen in the water," is a rather drab looking town with a population varying from 250 to 5,000, depending on the fishing seasons. But it is the location of the largest shrimp freezing and packing plant in the world.

Eduardo Rodríguez, plant superintendent for the Topolobampo Fishery, a Mexican corporation, says the shrimp bring the highest prices in the world market.

1 The word "Mayan" should be Mayo. The Mayo (Yoreme) people live in Sinaloa and their language is related to Yaqui and Tarahumara. The "Mayan" people are in southern Mexico. The word Topolobampo is of Mayo origin. Dr. Carman, in his report to Owen in 1881 (p.24) said that its meaning was "Hidden Water," but according to Anacleto Ramírez, a native Mayo speaker and translator of the Bible into his language, the first part of the word refers to a wild cat (*tigrillo*), perhaps an ocelot, with the place name ending for water (-bampo). The name Los Mochis is also of Mayo origin, and comes from the word for turtle. Also see Collard, *Vocabulario Mayo.*

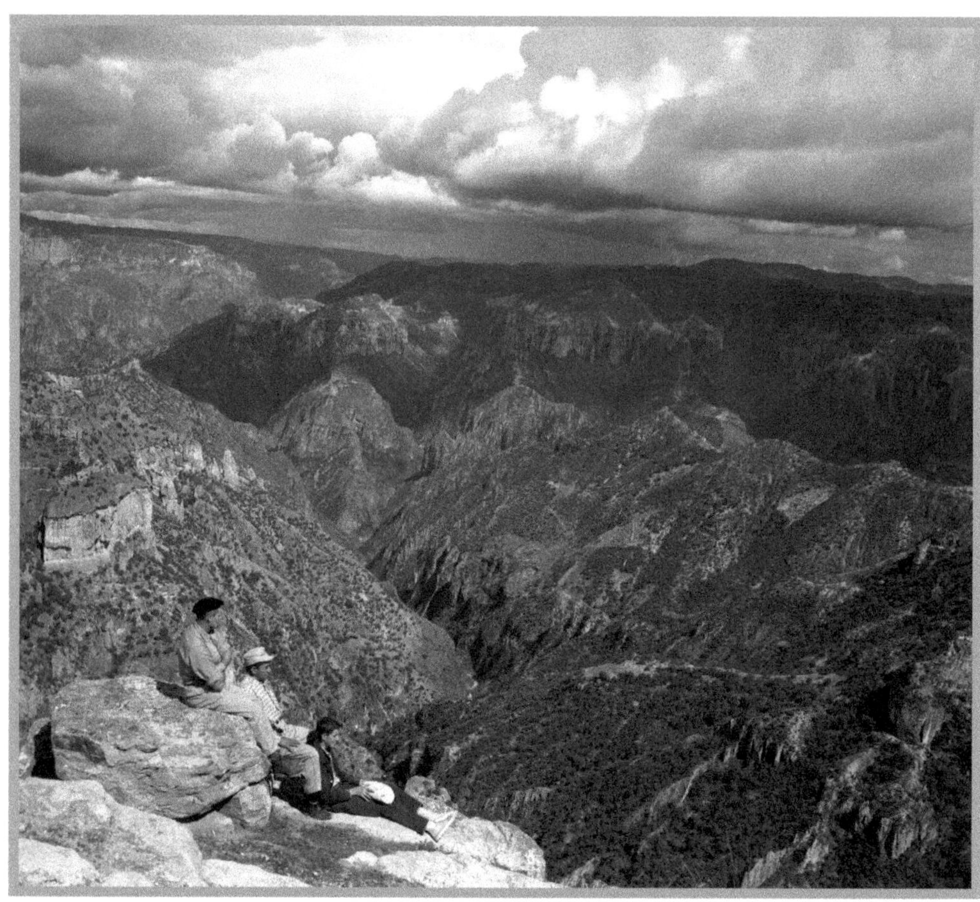

4.50. **SCENIC WONDER**—Steel rails pass within 100 yards of this observation point into the Barranca del Cobre. This scenic wonder is more than a mile deep (further downstream). From left: Ing. Streit, Ing. García Malo, and Joe Burgess, son of the photographer. Cleared areas where Tarahumaras plant corn can be seen down in the canyon (photo G.B. 1960).

The El Paso Times

- Sunday, October 16, 1960

New Era Of Trade, Commerce Seen For Mexico

Long Sought Rail Link to Open

By Glenn Burgess
Times Correspondent

Creel, Chih., Mexico.—A 60-year dream will be realized on Nov. 23.[1]

On that date a passenger train will leave from Juárez for Chihuahua City and a link will be formed that will place Mexico in a better position for trade and commerce.

At Chihuahua City the train will turn westward to the sea and go to Topolobampo on the Pacific Ocean. At the other end of the line will be Ojinaga, Chihuahua, right across the Rio Grande from Presidio in Texas.

That will make a complete line for Mexico from Kansas City to the Pacific.

President López Mateos, Chihuahua Governor Teófilo Borunda, many high ranking officials and newsmen will make the inaugural run.

The construction work is now in its final phase.

Rails manufactured by the Colorado Fuel and Iron Co. in Puebla, Colorado., and transported to Presidio, Texas by the Santa Fe, have been strung along practically all of the remaining 176 miles of grades between Creel, Chihuahua, and Hornillos, Sinaloa. This is the remaining link in the original 1,659-mile line from Kansas City to Topolobampo, that has defied construction men for 60 years.

SHORT SEGMENTS REMAIN

Three thousand men, using all sizes of modern machinery, are feverishly attacking the two remaining short segments of construction work on the line—17 miles in the upper Septentrión Canyon near the western border of Chihuahua, and seven miles in the Cuiteco Canyon, about midpoint in the 176-mile segment. An additional 3,000 men are preparing the grades for the actual track laying being done by a crew of 125.

1 This completion date was delayed by a year.

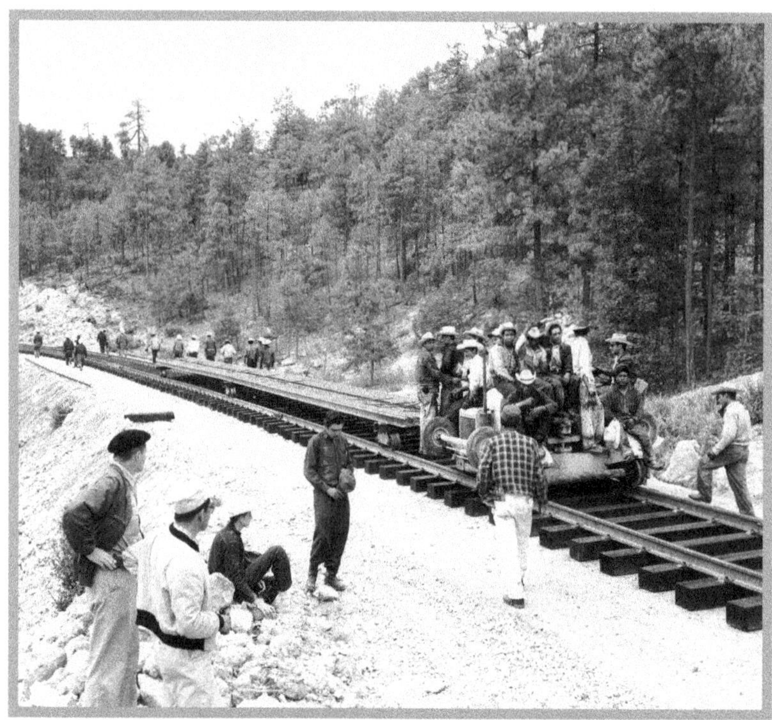

4.51. Moving rails down the line to where they will be connected to other rails (photo G.B. 1960).

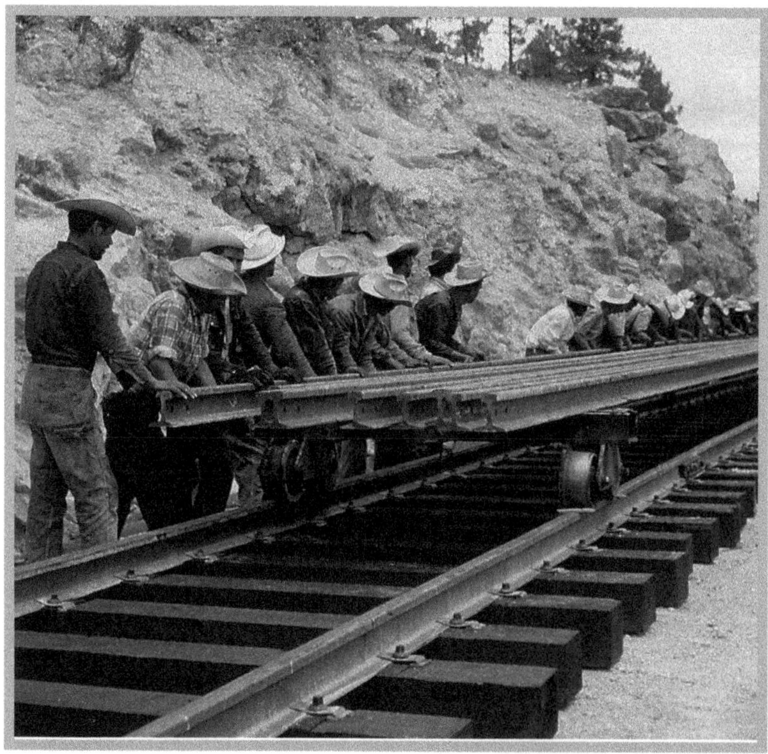

4.52. Getting ready to lift rails from the "truque" (photo G.B. 1960).

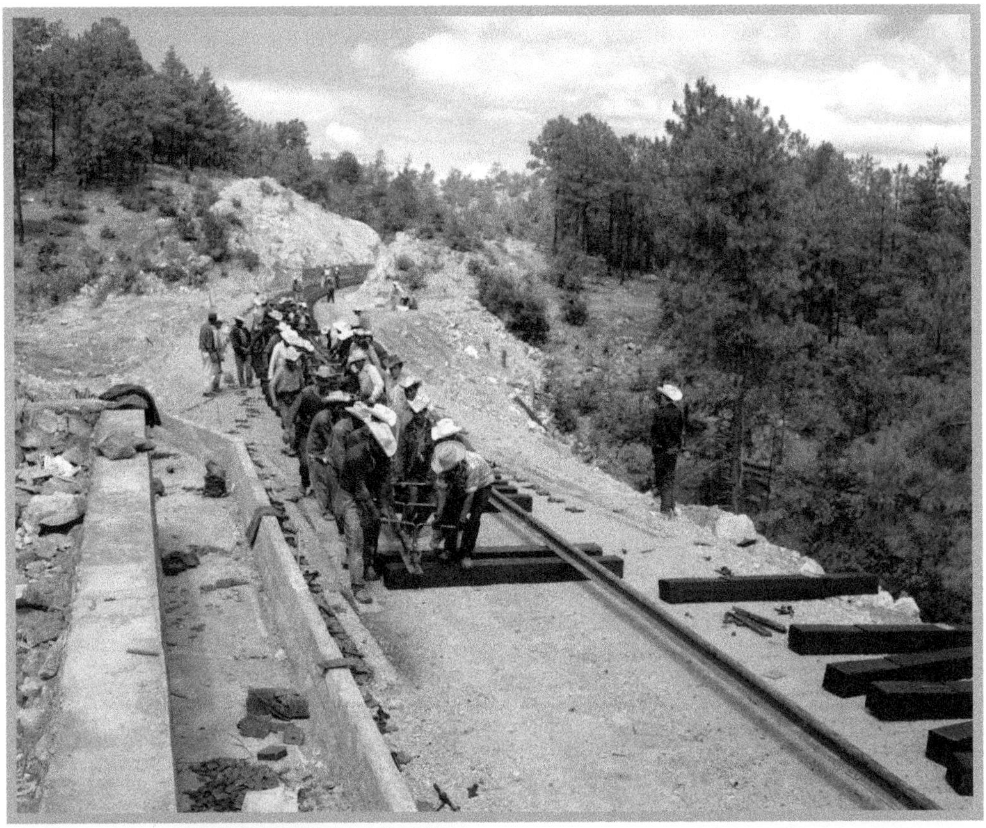

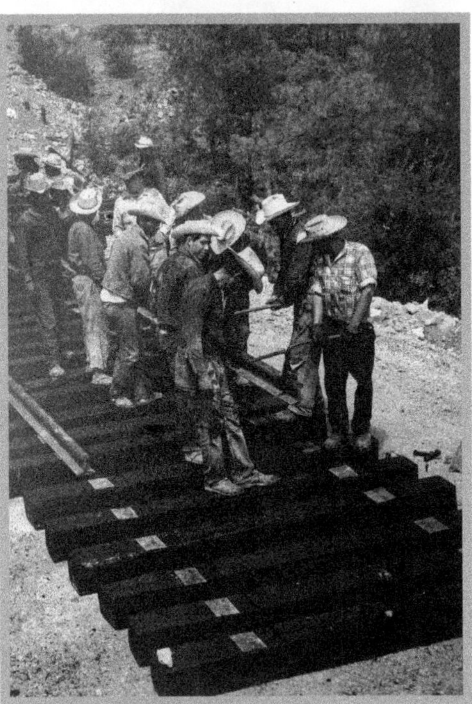

4.53 and 4.54. **HARD AT WORK—** This crew works in perfect symphony in the exciting task of laying the rails across the ties. Thirty-two men are required to lift the two 78-foot rails welded together. Welding process makes the rails far more stable (photos G.B. 1960).

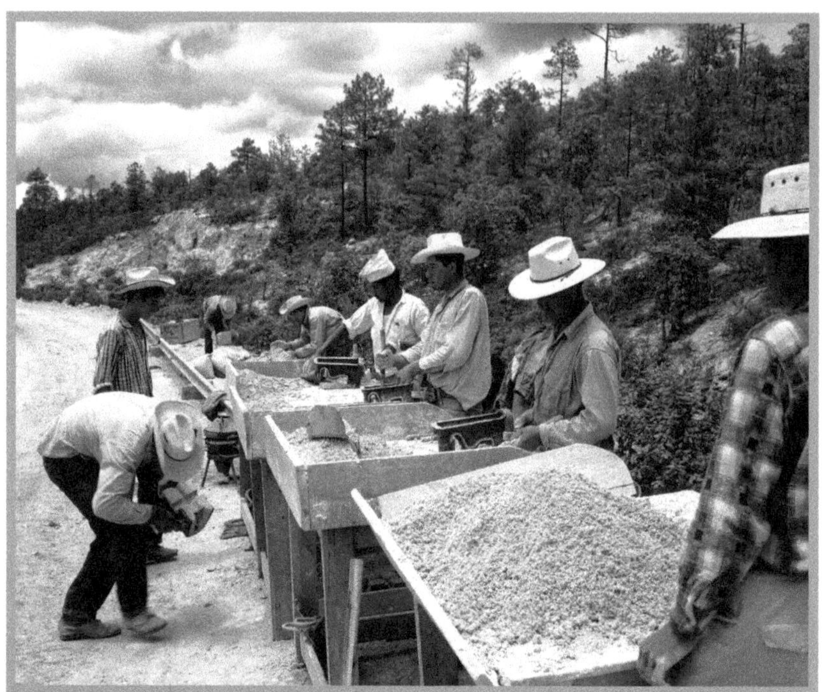

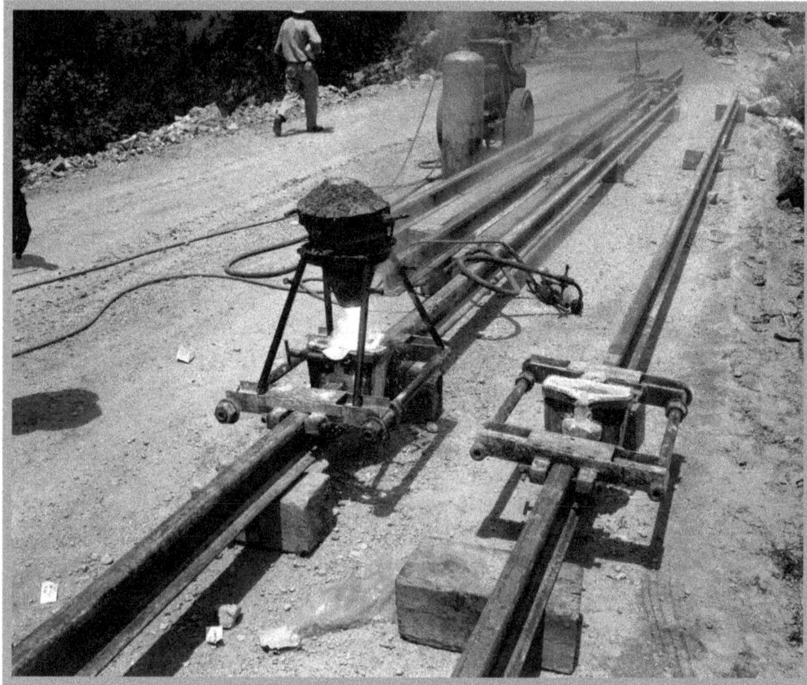

4.55 and **4.56.** For each rail connection to be welded, forms made of sand and clay are prepared at the site (upper photo). Then the rails are pre-heated to the desired temperature with petroleum burners, and the molten steel poured into the mold (photos G.B. 1960).

Finally, workmen in the woods of the high Sierra Madre are carving out 400,000 crossties[2] that are being creosoted at the plant at La Junta, while a concrete plant near San Blas, Sinaloa is manufacturing a new type of concrete crosstie for the 46-mile section between Hornillos and Chínipas on the western end.

While the work on the new line is rapidly nearing completion, the last step in the modernization of the 186-mile segment between Creel and Chihuahua City, the 4,100-foot Aguatos Tunnel through the Continental Divide at Creel [just north of Creel] lacks just 60 feet of rock removal for completion. This tunnel will eliminate some steep grades and 2.17 miles of track. New rails, crossties, ballast, better curves and new tunnels and much grade leveling has been completed on the entire route between Chihuahua City and Creel. At La Junta, where the Noroeste Railroad from El Paso and Ciudad Juárez connects with the Chihuahua al Pacífico, plans are being worked out for the modernization of the first 100 miles from La Junta to Madera.

2 For comments by one of the axe-men who made some of those ties, see the interview with Candelario López below.

All of the above adds up to a definite connecting of the rails between Creel and the West Coast by Nov. 23, the proposed date set for the President of Mexico to officiate at an opening ceremony. This is 1½ years earlier than the engineers on the job predicted in April 1959, when this writer went over the entire route.

DIFFICULT CHORE

Engineers today say it will be very difficult, but now possible to link up the rails by the November date.[3]

However, an unbelievable amount of construction work has been accomplished since April 1959.

The original dream of placing mid-continental United States 400 miles nearer the Pacific, conceived by A.E. Stilwell of Kansas City and now being put into execution by Mexico's top railroad construction engineer, Ing. Francisco Togno, will come true before another year.

The completed line will be extremely different from the railroads built by Stilwell in the early part of the century. Leading construction engineers consider the new line a complete

3 Engineers told me that they needed more time to get everything in top shape, but politicians entered the scene and forced an early completion.

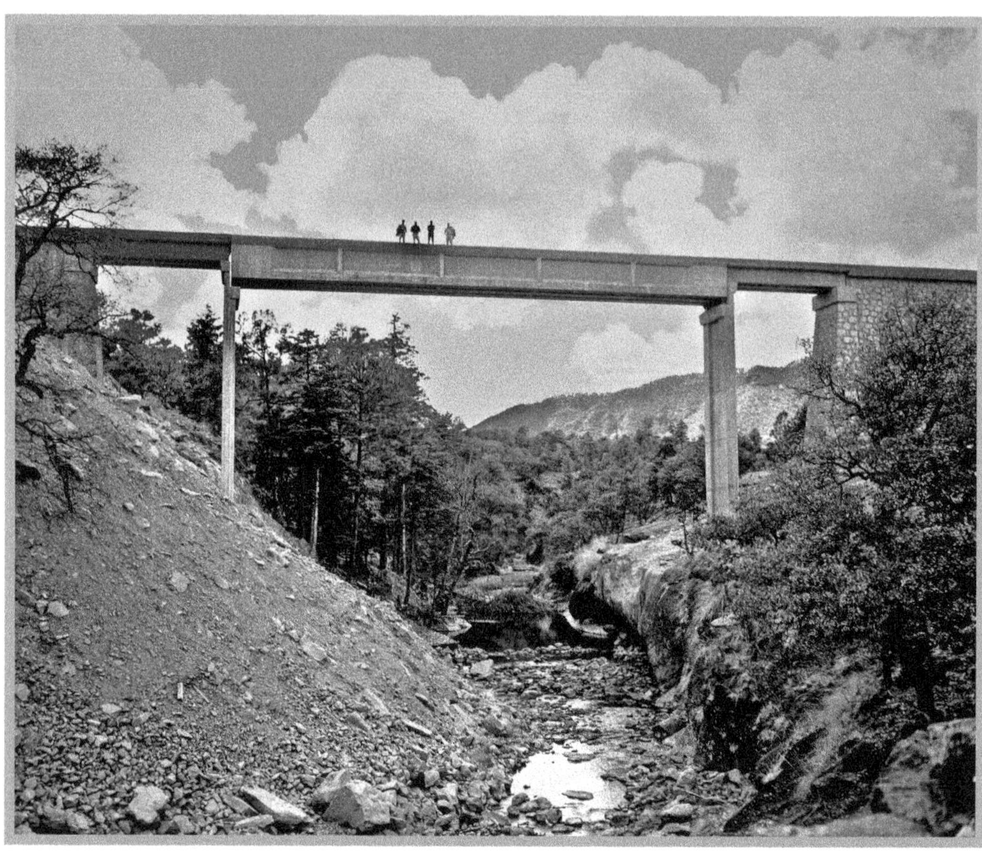

4.57. Engineers check the bridge at Cuiteco. The main beam of concrete is pre-stressed and has cables passing through tubes with a tension of 900 tons. This causes it to curve upwards eight or nine inches when no train is crossing (photo G.B. 1960).

revolution in railroad building. To maintain the required 2.5 per cent grade through the Sierra Madre, 84 tunnels, 35 bridges and four trestles have been built between Creel and Hornillos.

Two more long tunnels have been built between Creel and San Juanito on the original line, making a total of four between Creel and Chihuahua City. The tunnels of the entire line vary in length from a few yards to 1.2 miles. One bridge across the Rio Fuerte is approximately 100 feet high and 1,454 feet long. Still another, over the Rio Chínipas will elevate the trains almost 300 feet above the river. Yet another, near Cuiteco, is the only pre-stressed concrete railroad bridge in North America.[4]

Then, through the mountains and on most of the line between Chihuahua City to Topolobampo on the Gulf of California, the Mexican engineers have adopted the European system of plastic rail laying. On hand for the final track laying is France's expert, Ing. Michel Streit, a civil engineer who has specialized in new types of rail laying, especially in mountainous terrain. He is working closely with Mexico's competent track laying and rehabilitation construction engineer, Ing. Mariano García Malo, who is a combination electrical, mechanical and civil engineer.

Ing. García was in charge of the modernization of the old Southern Pacific Line between Nogales and Guadalajara, and of the more recent work between Creel and Chihuahua City.

The plastic system means setting the rail on a rubber plate, fastening the rail to the cross-tie with a large screw and a steel spring plate, setting the rail on the cross-tie at an inward angle of 1:40, and welding every other rail together, thus eliminating half of the conventional couplings.

FAST TYPE

Out on the level and straight stretches, they have welded three rails together, and in one section are experimenting with 800 yard-long welded rails. In the mountains, the originals are 78 feet long. With this system they can use the comparatively light 90-pound rail. In France, rail speeds of 203.36 m.p.h. have been attained with this type of construction.

Although 60 per cent of the distance between Creel and Hornillos is a curve, the engi-

4 Ing. Leal, an uncle of Francisco Togno, told me that the longer bridges all had this type of construction.

neers are preparing for an average train speed of 37 m.p.h. on the movement of freight. Ten Fairbanks-Morse locomotives of special design have been ordered at a cost of $2.4 million. Three will have steam heating facilities for passenger trains.

Preceding the laying of rails, the 3,000 workmen are doing everything possible to stabilize the grades, tunnel entrances, and walls of the cuts. This calls for grade leveling, construction of concrete and stone drainage structures and dislodging all predictable rock slides. The latter is very important because of the lack of visibility on curves. After the track is laid, good ballast rock, already prepared will be placed on the road-bed.

Laying the track is being done partially by hand. First, the rails were shipped by flat cars to Creel and then hauled by platform trucks and strung along the grade. Trucks keep ahead of the construction train and crews with the cross-ties. The cross-ties are put in place by hand and the 156-foot rail is carried to its location by a crew of 32 men armed with special tongs. Following the rail handlers is a crew of 21 men operating machinery designed to sink the large wood screws so that the spring clips will have the proper tension.

As the project moves forward, cost of transporting cross-ties becomes an important item, until in the western part of the State of Chihuahua, it becomes three times the original cost of the cross-tie. That is why they are using the concrete cross-tie in Sinaloa. Two blocks of reinforced concrete are held together by a piece of old rail. This tie, used extensively in France, has a life span of 50 years as compared to 30 years, the normal life of a creosoted wooden tie.[5]

TO HIT GAP

When the track-laying crew reached La Laja, approximately 55 miles west of Creel, a seven-mile gap in construction was encountered. Ing. García skipped past this and laid another additional 31 miles without benefit of a construction train. When he has finished this section, then he can come back to the seven-mile segment that should be ready for rails by that time. At present, the track is moving forward at the rate of 800 to 1,500 yards per day. The track laying began at Creel on June 1, and on Oct. 11, there were nearly 50 miles

5 For a discussion of wood and concrete ties, see "Justificación del uso del durmiente de concreto en los ferrocarriles Mexicanos" by Ing. Carlos Lezama G. in *Comunicaciones y Transportes*, pp.69-82.

of completed track.

Welding the rails together is a neat accomplishment. A movable steel foundry has to be set up at each connection desired. As in a large foundry, forms are made of sand and clay and attached to the connection. Then the rails are pre-heated to the desired temperature with petroleum burners, and the melted steel poured into the mold. This operation is delicate and is watched carefully.

The completed line will be an adaptation of the most modern methods of mountain penetration now being used in France, Switzerland, Germany and Sweden. The final 176 miles has been located, surveyed, designed and constructed under the direction of Mexican engineers. It is considered a nation-building development railroad financed entirely by appropriation from the Mexican government.

MAN STANDS OUT

Construction has been carried out by the Secretary of Public Works. One man stands out as the reason for completing the railroad. Ing. Francisco Togno started as a local contractor near Creel, then as the engineer in charge of locating the final route, the engineer in charge of the construction office at Chihuahua City, later as Chief of Construction of all railroads in Mexico, and now raised to the position of Assistant Minister of Public Works.

It is this engineer who convinced three presidents of Mexico that the project was feasible and that it should be completed.[6] He believed that the original promoter Stilwell used sound economic reasoning in proposing and initiating the railroad, and dedicated his life work to the fulfillment of the Stilwell dream.

Now, more is being said of making of Ojinaga, Topolobampo and possibly Ciudad Juárez, free ports so that goods shipped to and from the United States can be moved across Mexico in bond. Freight movement should require 20 hours from Topolobampo to Ojinaga and slightly longer from Topolobampo to Ciudad Juárez.

According to figures furnished by former Division Engineer Alfonso Rincón Benítez, who has handled most of the construction work, the rail dis-

6 In a letter to Glenn (June 17, 1962), Ing. Togno wrote that he had to deal with five presidents: Gen. Lázaro Cárdenas (1940); Gen. Manuel Ávila Camacho (1942); Lic. Miguel Alemán (1946), who suspended the work; Sr. Adolfo Ruíz Cortines, who attacked the work with great intensity; and Sr. Lic. Adolfo López Mateos (1959), who finished the work.

tance from Ojinaga to Topolobampo is 582 miles and from Ciudad Juárez to Topolobampo via the Noroeste Railroad, 661 miles. Stilwell figured the original distance from Kansas City to Topolobampo as 1659 miles. Mexican engineers believe they can effect considerable savings on Pacific freight to mid-continent cities in the U.S.

Studies are being made on passenger service and a special train from Texas is under consideration by the passenger department when the road gets into actual operation.

Now an international railroad promoted originally in Kansas City as the Kansas City, Mexico & Orient in 1900 may get into actual operation after 60 years. Other people thought of the line from El Paso and from Chihuahua City to the Pacific before Stilwell announced his plans.

At least three companies were involved in starting a railroad from Ciudad Juárez to La Junta via Nuevo Casas Grandes in 1897. By 1900, these same companies had built a line of 120 miles from Chihuahua City to Miñaca along the route between Chihuahua City and Topolobampo. The La Junta-El Paso segment was completed in 1911.

STARTED IN 1900

Stilwell organized his company early in 1900; started grading that year in Harper County, Kansas; laid his first rails at Emporia, Kansas on July 4, 1901; began construction from Chihuahua City towards Ojinaga in 1903; built the 62-mile Miñaca-Creel link in 1907; did much survey work in the Sierra, then lost his project in 1912.[7]

The receivers completed the line to Ojinaga late in 1928, the year the Santa Fe System purchased the entire line. In 1929 the Santa Fe sold the Mexico holdings to B.F. Johnston of Sinaloa and in 1930, completed their line from Alpine to Presidio. Johnston lost money on the railroad and passed his holdings on to the Mexican government in 1940.

Mexico began the location work on the final 176-mile segment in 1941. Some construction work was begun in 1953, and by 1955 it was moving along rapidly. Then in 1956 came larger appropriations and since that time every effort possible has been exerted to complete the line that is going

7 Stilwell died in 1917. Two weeks after he died, his wife jumped to her death from her twelfth story apartment in New York. She left a note saying: "I just could not live without my beloved one" (Kerr, p.105).

to cost in the neighborhood of $144 million.

When this line is finished, the rehabilitation work on the old Noroeste from La Junta to El Paso should begin. Considerable survey work has been done already on the first 100 miles between La Junta and Madera.

This project now takes on interest for two sections in the United States.

EXTENDED COVERAGE

Coverage of the inaugural trip will be one of the most extensive in Mexico.

To date 34 newsmen, radio and TV crews have filed for reservations. Included are the *El Paso Times, Denver Post, Fort Worth Star Telegram, National Geographic Magazine, Time* and *Life* magazines, *Arizona Republic, Mexico City* and *Chihuahua City, Los Angeles Times* and many others.

A special bus will take the newsmen from Juárez to Chihuahua City where they will board the train.

After returning to Chihuahua City Governor Borunda will be host at a barbecue on his ranch.

President Mateos summed up the opening of the rail link by saying: "Mexico takes another positive step forward in its economic and commercial development."

A "Truque" Accident

The rails were sometimes moved by laying rails across a set of wheels known as a "truque" and then moved by a motorized car (called "pas pas" for the sound they made) along the already laid rails to where they were working.

Salvador "Chava" Bustillos, who worked on one of these crews as a young man, tells that in 1960, when they were working just up from El Lazo, a "truque" with four rails got loose from the motor car and started rolling down the grade, picking up speed as it went. Stopped in the tunnel just before El Lazo were two new motorized, canvas-covered cars which carried people, and the "truque" ran into the cars, killing four people and wounding two others.

Antonio González, who was the "maquinista" (driver of the railroad motor car) for Mariano García Malo told me that his brother was on the scene when that happened and he told him that small rocks had been placed behind the wheels of the "truque" to keep it from rolling down the rails while it was disconnected from the motorized car which was used to pull it. But then a huge thunder-clap shook the whole area and the rocks fell off the rails and the "truque" started moving. His brother tried to stop it, but could not. The motorized car which pulled the "truque" then went along behind honking its horn to alert people, but the people in the tunnel did not realize what was happening.

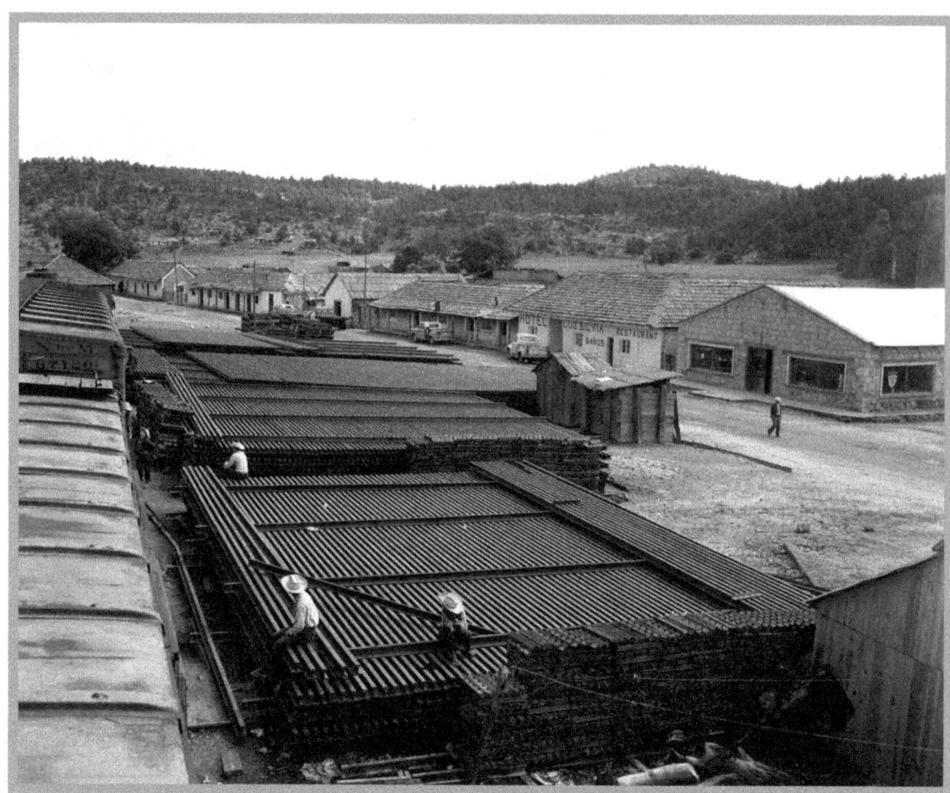

4.58. Rails manufactured at CF&I's Pueblo (Colorado) Plant were shipped via the Santa Fe and entered Mexico at Presidio, Texas. Stacked at Creel, the rails were hauled southwest along 176-mile roadbed by platform trucks and strung along grade (photo G.B. 1960). [The stone building was the local gasoline station. Now it is the Hotel Cascada Inn, owned by the same family. To the left is the Hotel Luz Silvia, now the Hotel Parador de la Montaña.]

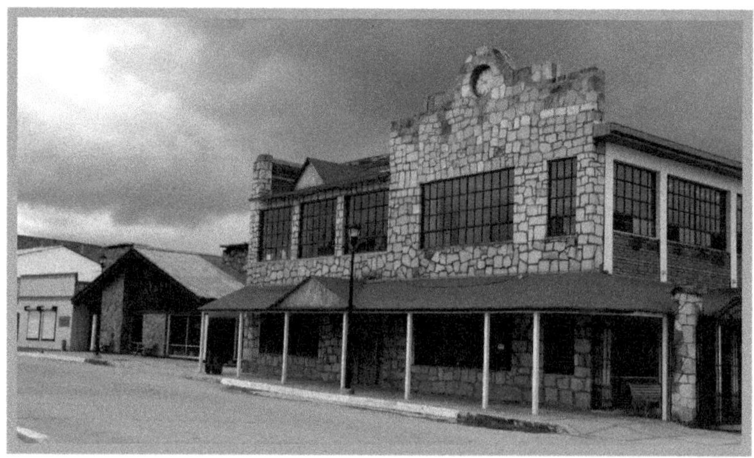

4.59. The same buildings in 2012 (photo D.B.).

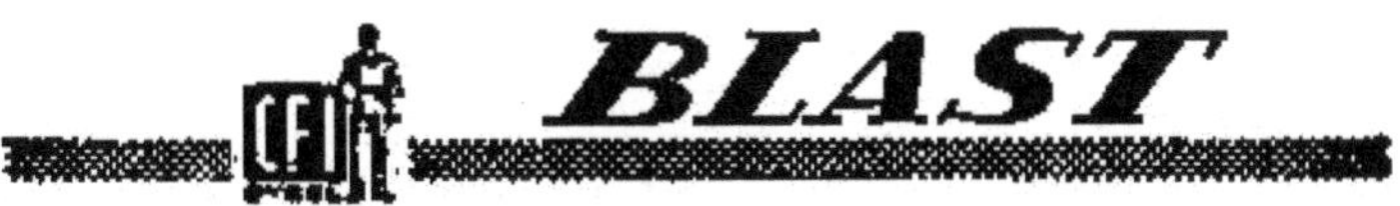

- Monday, May 22, 1961

Sierra Challenge

Men move mountains to lay 176 miles of CF&I rails, close Kansas-Mexico's west coast gap

By Glenn Burgess[1]

CHIHUAHUA, Mexico—The treasure of the Sierra Madre is no longer gold. It is steel—CF&I steel which is helping the Mexican government complete a fantastic railroad project that will link Mexico's west coast with Kansas City, 1,569 miles away.

For the past five years 6,000 men have been fighting nature with modern machinery and bare hands so that 176 miles of CF&I rails could be laid between Hornillos and Creel, right through the forbidding Sierra Madre mountains. And within a few months, it will be possible to haul freight and maybe transport passengers from El Paso and from Presidio, both Texas border towns, to Topolobampo, right on the Gulf of California.

Originally, the Creel-Hornillos line was scheduled to be completed in November of 1960, but unbelievable difficulties in building bridges and cutting tunnels through the Sierra Madre mountains prevented the construction crews from meeting the deadline. A target date of November 1, 1961, has now been set, which would give engineers enough time to level the track and stabilize the grades after the rainy season.

Although at least three companies were involved in starting a railroad from Ciudad Juárez, across from El Paso, to La Junta via Nuevo Casas Grandes in 1879, the dream of placing mid-continental United States 400 miles nearer the Pacific was conceived originally by A.E. Stilwell, of Kansas City, some 61 years ago.

Organized as the Kansas City, Mexico & Orient early in 1900, Stilwell's company started grading that year in Harper

1 The *CF&I Blast*: a bi-weekly newspaper for the Colorado Fuel and Iron Corporation, Pueblo, Colorado.

County, Kansas. The first rails were laid at Emporia, Kansas, on July 4, 1901, and in 1903 Stilwell began construction from Chihuahua City towards Ojinaga, building the 62-mile Miñaca-Creel link in 1907. In 1912, after he had done extensive survey work in the Sierra, Stilwell lost the project.

The receivers completed the line to Ojinaga late in 1928, the year the Santa Fe purchased the entire line. In 1929 the Santa Fe sold the Mexico holdings to B.F. Johnston, of Sinaloa and in 1930 completed their line from Alpine to Presidio. Johnston, however, lost money on the railroad he had bought and passed his holdings on to the Mexican government in 1940.

Location work on the final 176-mile segment between Creel and Hornillos was started in 1941 by Mexico and some construction was done in 1953. By 1955 the project was moving along rapidly. Then, in 1956, came larger appropriations and since then no efforts have been spared to complete the $144 million job as quickly as the rugged terrain and Mexico's rainy season will allow.

Few railroad projects anywhere in the world will have required such a display of engineering know-how and sheer human tenacity as the Hornillos-Creel link, which was located, surveyed, designed and constructed under the direction of Mexican technicians.

Led by Mexico's top railroad construction engineer, Ing. Francisco Togno, the 6,000 workers on the project have built 84 tunnels, 35 bridges and 4 trestles to maintain a 2.5 per cent grade through the Sierra Madre. Of these 6,000 men, half are blasting rock, dislodging mammoth boulders, cutting tunnels and pouring concrete, while the other half of the work force is preparing the grade for the actual track laying.

In addition, workmen in the woods of the high Sierra Madre have turned out 400,000 crossties to be creosoted at a plant in La Junta, and a concrete plant near San Blas is manufacturing a new type of concrete cross-tie for the 46-mile section between Hornillos and Chínipas, on the western end of the line.

The CF&I rails used on the project were manufactured at the Pueblo Plant and transported to Presidio by the Santa Fe. From the border the rails were shipped to Creel on flat cars, then hauled by platform trucks and strung along the grade.

Whenever it is feasible to eliminate conventional couplings, the rails are welded end to end with a portable steel foundry. On the level and straight stretches, three or four welded rails are used, and on one section of the new line the Mexican engineers are experimenting with 2,400-foot welded rails. In the mountains, twists and bends limit the length of the rails to 78 feet (two welded rails).

It takes 32 men to place the rails upon the cross-ties and another crew of 21 to operate the machinery that fastens the rails to the cross-ties. All told, the track laying crew totals 125.

As the project moves forward, the cost of transporting the cross-ties goes up, until in the western part of the State of Chihuahua it becomes three times the original cost of the cross-ties. For this reason, only concrete cross-ties, made of two blocks of reinforced concrete held together with a length of rail, are used in Sinaloa. This type of tie, used exclusively in France, has a life span of 50 years, as compared to 30 years for a creosoted wooden tie.

Linking Creel and Hornillos is not, however, the only construction task undertaken by the Mexican government in order to make its railroad network one of the finest transportation instruments in the Western Hemisphere. From Creel to Chihuahua City, the existing 186-mile segment is being completely modernized and some 2.17 miles of steep grades will be eliminated by the construction of the 4,100-foot Aguatos Tunnel through the Continental Divide at Creel. And at La Junta, where the Noroeste Railroad from El Paso and Ciudad Juárez connects with the Chihuahua al Pacífico, work has started to renovate the first 100 miles toward Madera, on the La Junta-Juárez line. In these projects, too, CF&I rails will be instrumental in placing Mexico in a better position for trade and commerce.

Already there are talks of making Topolobampo, on the Gulf of California, and Ojinaga and Juárez, across from Texas, free ports so that goods shipped to and from the United States can be moved across in bond (that is, under care of bonded agencies until duties or taxes are paid). Freight movement would require 20 hours from Topolobampo to Ojinaga (582 miles) and slightly longer from Topolobampo to Juárez (661 miles). Studies are also being made on passenger service and a special train from Texas is under consideration by the passenger department of the Mexican National Railroads

(N de M) when the line gets into operation.

Not only a new and significant factor in the economical development of Mexico, the Creel-Hornillos link will stand as a monument to railroading and to the men who have literally moved mountains to build it. And for the CF&I steelworkers who made the rails needed for this engineering and construction achievement, there is much pride to be had in knowing that the product they've turned out is helping a friendly nation on the road to progress.

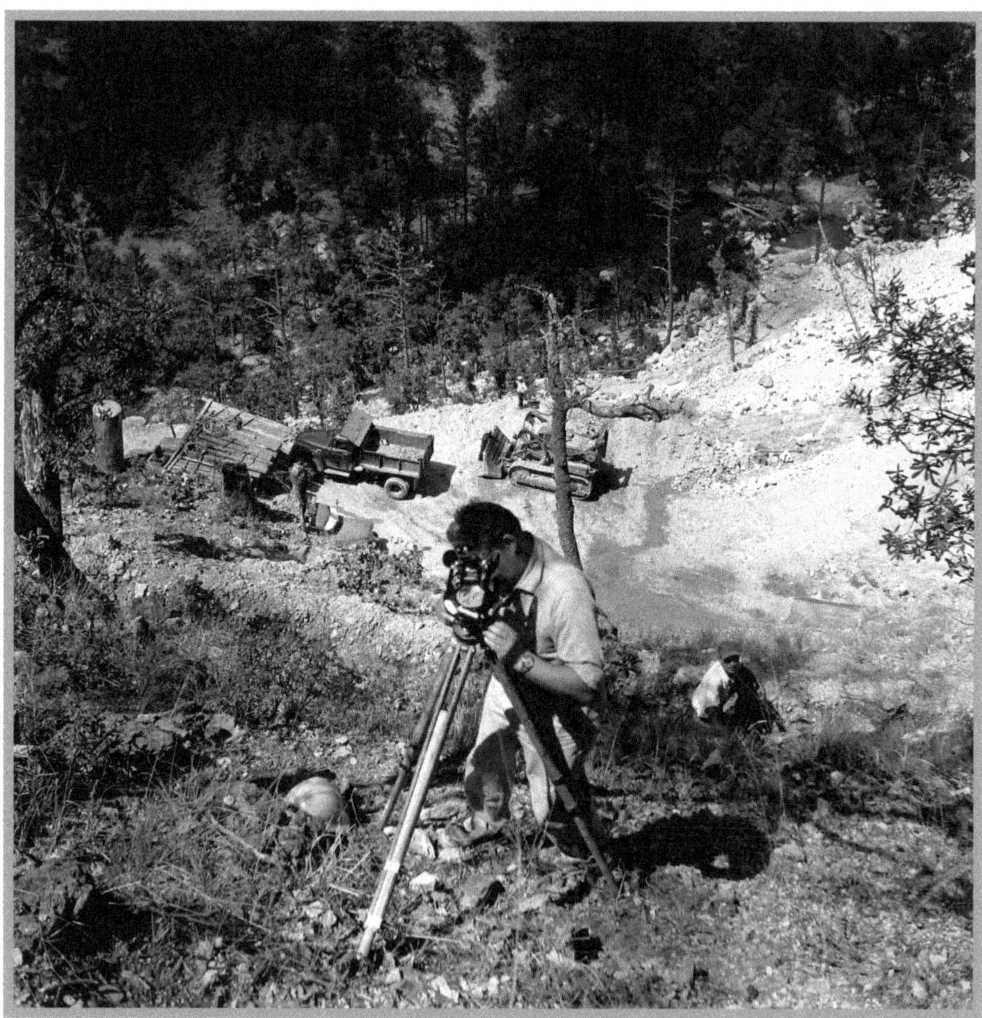

4.60. Engineer on a steep slope, surveying for a tunnel near Cuiteco (photo D.B. 1959).

Engineers and Their Families

Given the duration of the construction of the Chihuahua al Pacífico Railroad, family life continued in the Sierra. Many of the engineers' families lived with them in the camps. School teachers were provided by the Secretary of Public Education (SEP) with their salaries being paid by the Department of Communications and Public Works (SCOP). Engineer Miguel Leal married one of the local teachers at the Areponápuchi camp, Ernestina Quezada Villalobos, who was from Creel. They have been married more than 50 years. Other engineers chose to have their families in Chihuahua City.

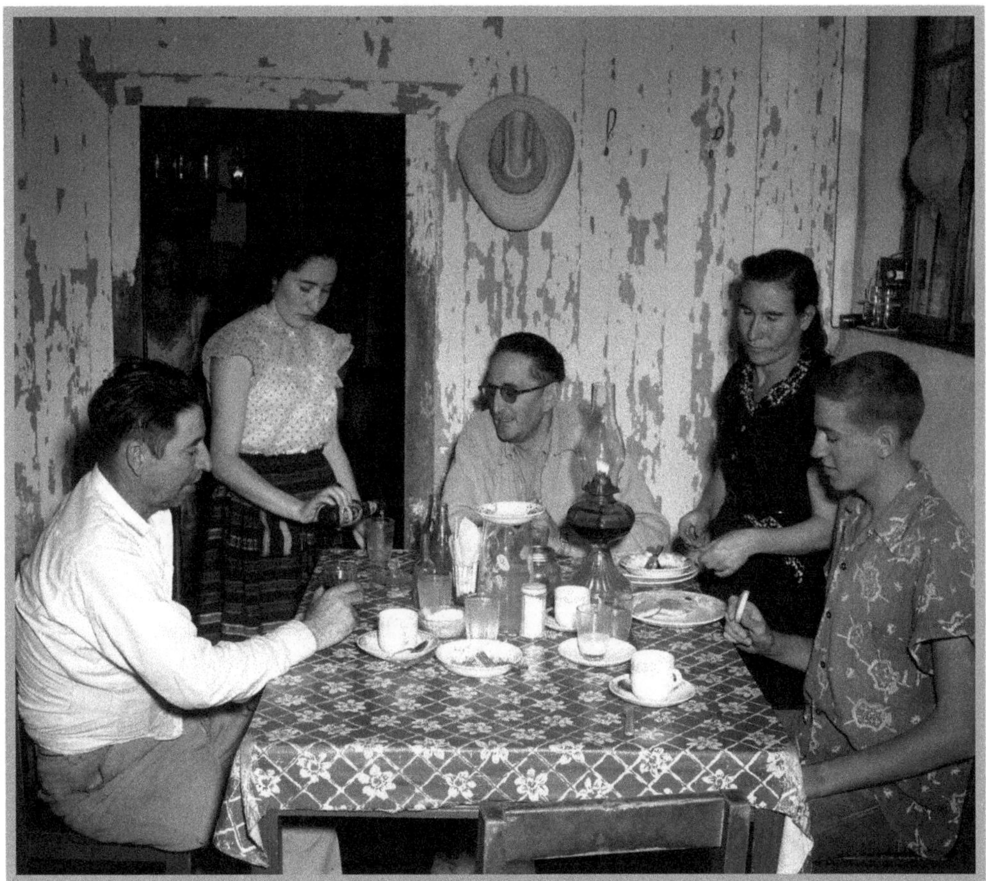

4.61. Ing. Jorge Togno in the center with Don Burgess on the right. The others are unidentified (photo G.B. 1955).

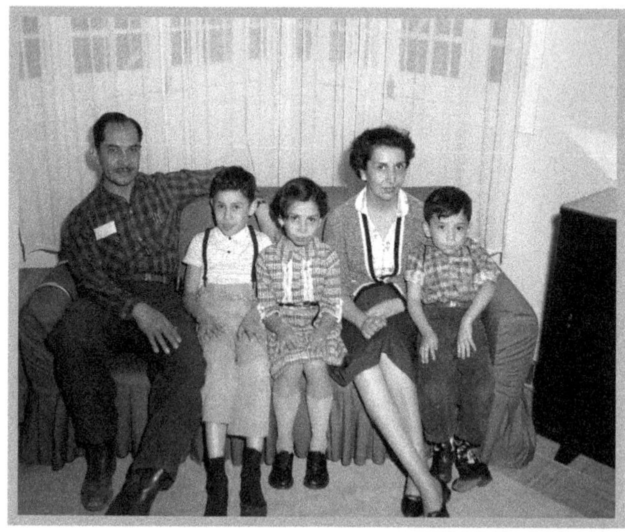

4.62. Ing. García Malo with his wife and three children in their Chihuahua City home (photo G.B. 1960).

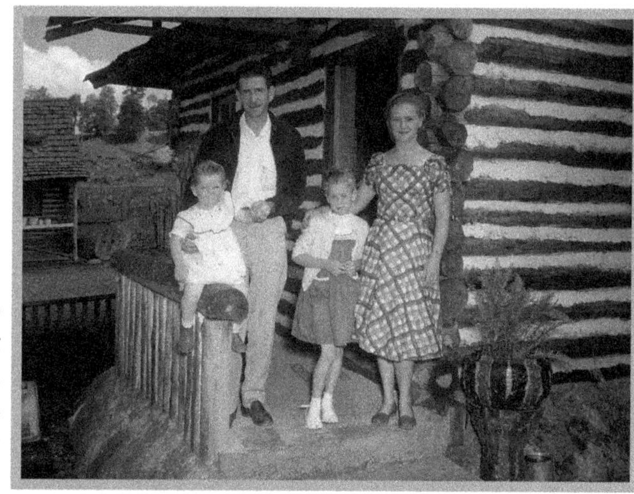

4.63. Ing. Ramón Togno Purón, younger brother of Francisco Togno, with his wife Guita and two daughters (photo G.B. 1955).

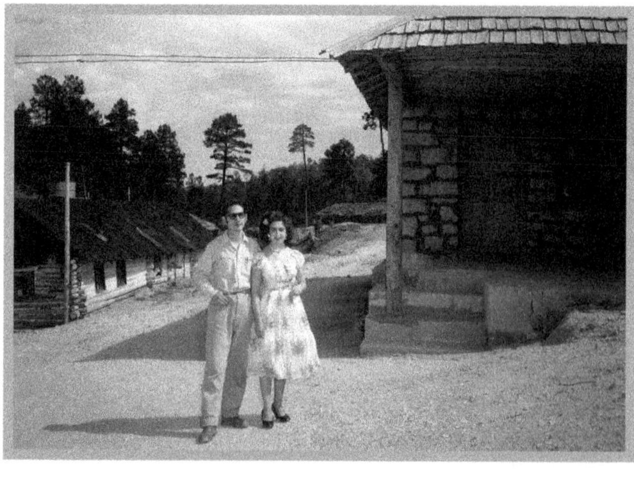

4.64. Ing. Arturo Múzquiz Orendein (graduate of Politécnico Nacional) with his wife in Areponápuchi (photo G.B. 1955).

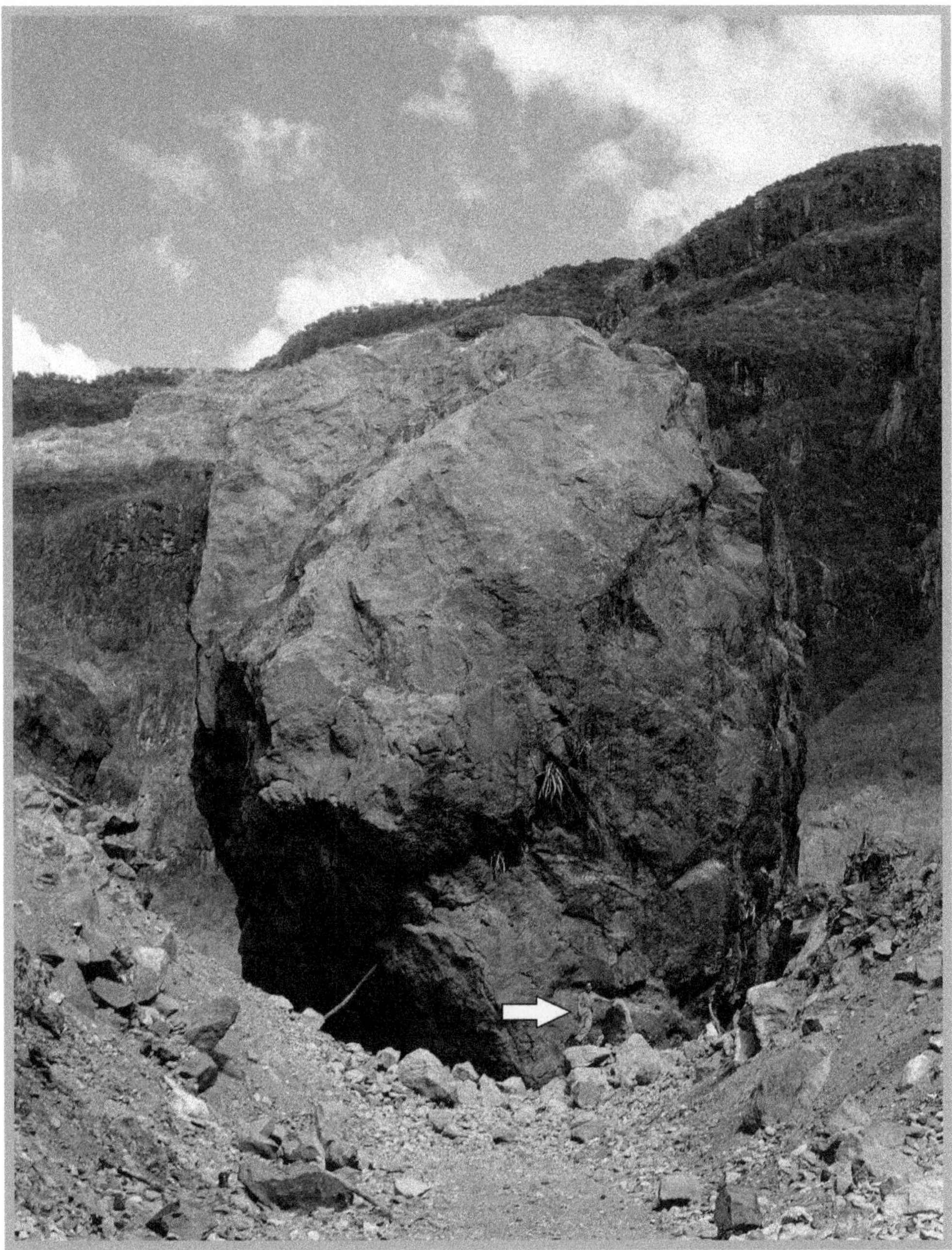

4.65. A formidable obstacle during construction of the Creel-Hornillos line in Mexico, this 9,000-ton, 75-feet high rock rolled down onto the grade after attempts had been made to stabilize it. The man standing at the base of the rock is Mariano García Malo, a Mexican track laying and rehabilitation construction engineer working on the huge project. To open way for the 125-man crew laying 176 miles of CF&I rails, 6,000 workers had to blast a path through the Sierra Madre. (photo G.B. 1960).

The Inauguration

news article by Ramon Villalobos
with notes by Don Burgess

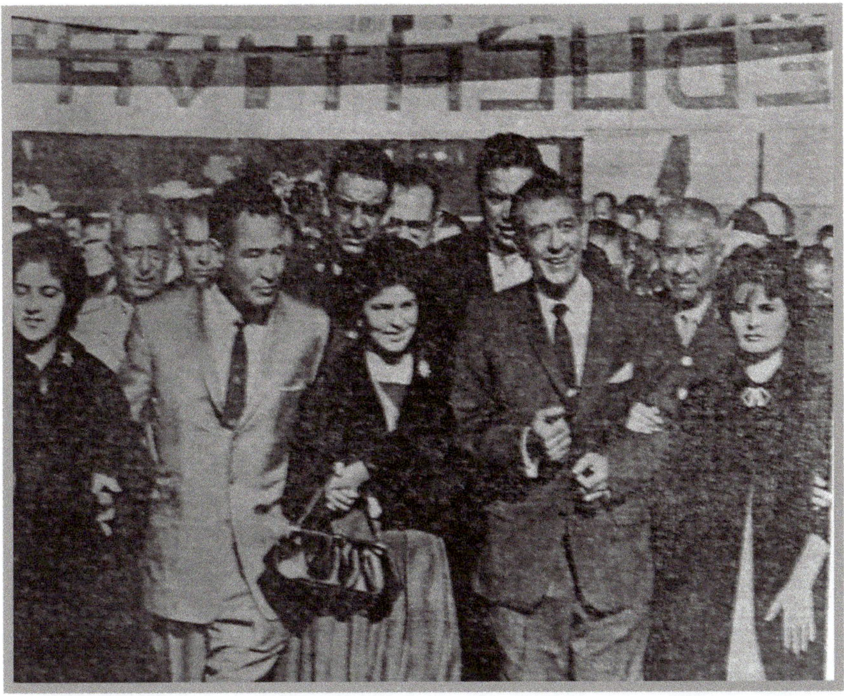

5.1. SMILING PRESIDENT AND CHARMING GREETERS – Each time President Adolfo López Mateos alighted from his Presidential train on the way to inaugurate the Chihuahua al Pacífico Railroad, he was greeted by cheering, hat-waving crowds. Here the president is shown with Chihuahua Gov. Teófilo Borunda, flanked by three charming señoritas at Anáhuac where the president dedicated a Social Security hospital. He also dedicated other public works en route including two dams (photo by Jorge Bate, *El Paso Times*).

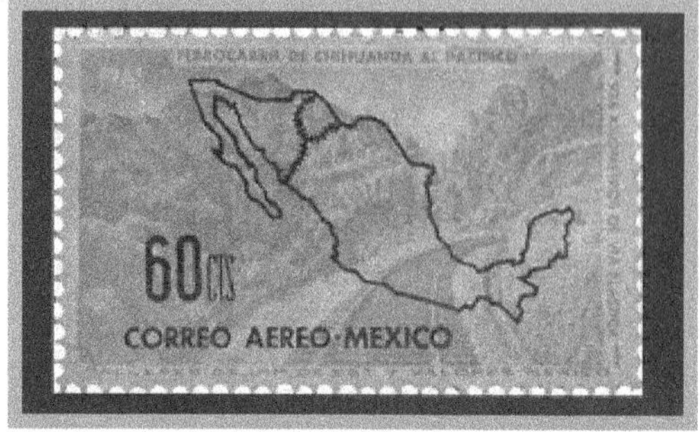

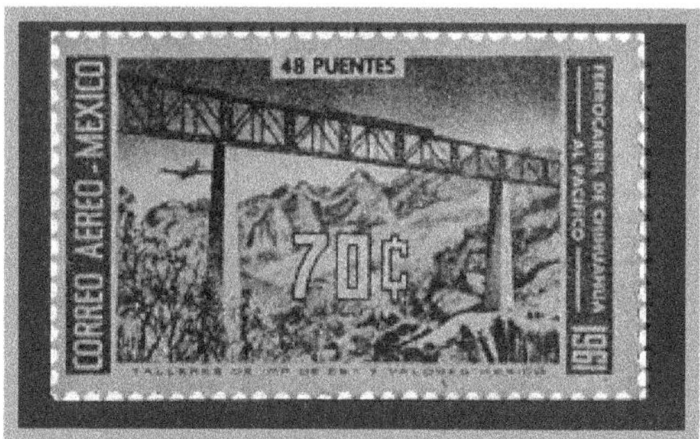

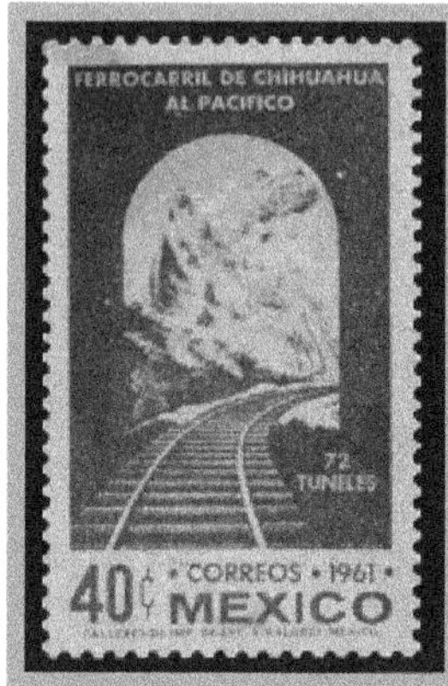

5.2, **5.3**, and **5.4.** Commemorative Stamps of 1961.

Commemorative Stamps

Three commemorative stamps were issued for the inauguration of the train (four million of each). One stamp shows the route, the second shows a tunnel, and the third shows a plane flying under a railroad bridge. This last stamp illustrates an event that took place at the Chínipas Bridge. Ing. Leal tells how Captain Fierro of Los Mochis flew under the Chínipas Bridge shortly after it was finished, a dangerous act because of some cables under the bridge. As a result, he had his pilot license taken away. But then officials in Mexico City heard what he had done. They told him that, if he would fly under the bridge again, so that photographs could be taken for using on a stamp (a movie was also taken), then his license would be reinstated. This is what happened.

The El Paso Times

- November 23, 1961

President Opens Rail Festivities

by Ramón Villalobos[1]

Chihuahua City, Chih.— President Adolfo López Mateos of Mexico arrived here by plane Wednesday evening with all but two of his cabinet members and was greeted by a roaring ovation from persons lined up all the way from the airport to downtown Chihuahua City.

The president is here to inaugurate the opening of one of the biggest projects ever attempted by the country, the Chihuahua al Pacífico Railroad which begins at Ojinaga and ends at Topolobampo on the Pacific Coast.

For the first time in the history of Mexico, the president's cabinet accompanied him on a trip of this kind. The entire cabinet, except the ministers of Foreign Affairs and War, arrived in the city with López Mateos.

Other high national and state officials who arrived here Wednesday afternoon included 36 senators, 36 congressmen and 66 municipal mayors from the state of Chihuahua.

The president was driven in an open car through the cheering throngs of people after his arrival to the municipal palace in downtown Chihuahua City.

The governor of Chihuahua, Teófilo Borunda, called a special session of the state legislature when the president arrived to give Mateos the special achievement award that was recently created by the legislature.

Mateos is the first president to receive the medal which was presented to him by Congressman Félix Cervantes of Juárez, president of the Chihuahua legislature.

A banquet was held in the Chihuahua City municipal gymnasium Wednesday night in the president's honor, with 700 invited guests attending.

Mateos and his party and newsmen will leave at 7 a.m. Thursday aboard the special Presidential train for Santa Barbara, Chihuahua, where he officially will inaugurate the railroad.

1 For some unknown reason, Glenn did not attend or write about the inauguration.

The Inauguration on November 24, 1961

The actual inauguration took place at Santa Barbara, near the village of Témoris, where the train does a complete horseshoe curve inside a tunnel, goes underneath a waterfall, across several bridges, etc.

Ing. José Eloy Yáñez Bordier, who attended the ceremony at the invitation of Ing. Mariano García Malo, wrote the following:

> I had the luck to be at the open space where the flag pole was located next to the station. The presidential train going south arrived, the platform of the last car stopping in front of the flag pole. President Adolfo López Mateos stepped down, saluted the national flag accompanied by the corresponding military honors, and returned to the tracks, where Ing. Lira Arciniega put in the president's hands the last screw, especially prepared for this ceremony with a golden head. Then he took the handles of the machine that twists the screws into the railroad ties and put into its place the final screw, symbolically concluding the construction of the Chihuahua al Pacífico Railroad. Then he went with those accompanying him to the unveiling of the large monumental plaque.

The huge plaque, made of old railroad rails, stated:

> This work was put into operation by Adolfo López Mateos, Constitutional President of the Republic, in commemoration of the fiftieth anniversary of the Mexican Revolution. Investment up to 1958 $390,000,000.00 (pesos). Investment from 1959-1961 $744,000,000.00. Secretary of Public Works.

Then the president stated:

> Today, the 24th of November of 1961, I am honored to turn over to the people of Mexico this work which represents the continued effort of many generations and above all the genius and the creative capacity of the Mexican.

And the almost impossible dream became a reality. The challenge of the sierra had been met.

Names of Railroad Stops and Sidings

It can be seen that some stops were named after already existing towns such as Chihuahua (a mining town, the meaning of whose name is disputed) and El Fuerte (the site of an old fort).

Some of the railroad towns have indigenous place-names such as Teméichi (Tortilla Place), Cuitéco (Throat), and Bahuichibo (Misty Plain).

Others have Spanish names such as San Juanito, a town which began as a railroad center and was named after a nearby ranch. The Spanish name Pitorreal is the translation of the name of the original Tarahumara ranch called Reláibo, "Plain of the Imperial Woodpecker". (This woodpecker, the largest in the world, is apparently now extinct. I saw one about 1970.)

Cuauhtémoc was originally a ranch called San Antonio de Arenales, but in 1927 the town that grew up there was given the name of the last Aztec Emperor. Santa Bárbara, as the Témoris station is sometimes called, is the name of a mine located nearby.

Some stops and sidings were named after people who had great interest in the railroad such as governors of Chihuahua (Enrique Creel, José María Sanchez, etc.), business men such as Julio Ornelas, and railroad personnel such as Francisco M. Togno, Ulises Irigoyen, etc.

Some names derive from the geography. El Divisadero refers to an overlook where the train stops for people to see the Copper Canyon. The Tarahumara word for overlook is "rekuata".

La Junta (The Joining Place) was so named since the Noroeste Railroad and the Chihuahua al Pacífico came together at that point. (See the text for explanations of other names.)

CHIHUAHUA PACIFIC RAILWAY CO.

Passenger Service.

CONDENSED PASSENGER TRAIN SCHEDULE EFFECTIVE MAY 15, 1962.	
TRAINS SOUTHBOUND TUESDAY AND FRIDAY	TRAINS NORTHBOUND MONDAY AND THURSDAY
0Ks. Lve 3:00PM.OJINAGA	Arr 6:00AM. 921 Ks
268 " Arr 9:00 "CHIHUAHUA	Lve 12:01 " 653 "
Lve 9:30 "	Arr11:30PM
883 " Arr 2:25 " SAN BLAS	Lve 7:00AM 38 "
Lve 2:35 "	Arr 6:50 "
921 " Arr 3:30 " LOS MOCHIS	LVE 6:00 " 0 "

RAIL AND PULLMAN FARES.

	FIRST CLASS	RECLINING SEATCOACH	LOWER BERTH	UPPER BERTH	DRAWING ROOM	COMPARTMENT
OJINAGA TO SAN BLAS...	$ 8.91	$ 9.97	$ 4.23	$ 3.38	$ 16.06	$ 11.83
OJINAGA TO LOS MOCHIS...	$ 9.32	$10.42	$ 4.42	$ 3.54	$ 16.78	$ 12.37
CHIHUAHUA TO SAN BLAS.....	$ 6.23	$ 6.97	$ 2.96	$ 2.36	$ 11.22	$ 8.27
CHIHUAHUA TO LOS MOCHIS..	$ 6:64	$ 7.42	$ 3:15	$ 2.52	$ 11.95	$ 8.81

Note: (1).-All Trains run on Mexico's "Hora del Centro", which is the same as Central Standard Time in the United States.

(2).-To Occupy Pullman accommodations, the purchase of First Class Ticket is required. Minimums: Compartment - 2 ½ First Class Tickets
Drawing Room- 3 First Class Tickets.

(3).-Equipment: Sleeper - 10 Sections, 2 Compartments, 1 Drawing Room.
Dining Car - (between Ojinaga and Los Mochis).
Reclining Seats Coach - (between Ojinaga and Los Mochis).

(4).-Railroad Tickets and Pullman Fares are quoted in U.S.Cy. at $12.50 Rate of exchange.

(5).-No responsibility is assumed for errors in this Time Table, inconvenience or damage resulting from delayed trains or failure to make - connections.

(6).-Schedules and rates herein are subject to change without notice.

CHIHUAHUA PACIFIC RAILWAY COMPANY
General Freight and Passenger Agency
TIME TABLE PASSENGER SERVICE
OJINAGA, CHIH.- LOS MOCHIS, SIN.

MLS	TRAINS SOUTH BOUND (READ DOWN) LEAVE: TUESDAY AND FRIDAY		TRAINS NORTH BOUND (Read Up) ARRIVE: TUESDAY AND FRIDAY
0	3:00 PM	OJINAGA	6:00 AM
19	3:33 "	CHAPO	5:15 "
27	3:50 "	LA MULA	4:54 "
35	----	LAGUNITAS	----
46	----	VOLCANES	----
50	----	VINATAS	----
57	4:51 PM	CHILICOTE	3:48 AM
71	5:18 "	PULPITO	3:14 "
80	5:35 "	FALOMIR	2:55 "
87	5:55 "	PICACHOS	2:37 "
95	6:15 "	SAN SOSTENES	2:19 "
103	6:36 "	COLONIAS	2:05 "
109	6:49 "	ENCANNTADA	1:52 "
120	7:07 "	MORREON	1:34 "
131	7:28 "	TRANCAS	1:14 "
147	7:57 "	ALDAMA	12:45 "
152	----	CALERA	----
158	8:19 "	MULLER	12:26 "
163	8:34 "	TABALAOPA	12:15 "
	ARR: 9:00 "		Lve: 12:01 "
167		CHIHUAHUA	
	Lve: 9:30 "		Arr: 11:30 PM
176	9:49 "	FRESNO	11:09 "
183	10:01 "	SALAS	10:55 "
189	10:10 "	PALOMAS	10:46 "
194	----	ARATZA	----
200	10:33 "	SANTA ISABEL	10:24 "
203	----	BAEZA	----
211	10:52 "	CHAVARRIA	10:06 "
217	11:06 "	SAN ANDRES	9:52 "
229	11:36 "	MESA	9:23 "
233	----	BUSTILLOS	----
237	11:48 "	ANAHUAC	9:08 "
249	12:06 AM	CUAUHTEMOC	8:51 "
257	12:21 "	CASA COLORADA	8:37 "
263	----	CIMA	----
265	12:36 "	PEDERNALES	8:21 "
272	----	ROSARIO	----
	ARR: 1:05 "		Lve: 7:55 "
281		LA JUNTA	
	Lve: 1:35 "		Arr: 7:25 "
287	1:47 "	MIÑACA	7:10 "
293	----	LA UNION	----
296	2:12 "	GONZALEZ	6:45
			-------→
302	2:27 AM	TERRERO	6:31 PM
309	----	SIGOYNA	----
313	3:04 "	PICHACHIC	5:55 "
319	3:23 "	ATAROS	5:35 "
326	3:42 "	TREVIÑO	5:16 "
329	----	TALAYOTES	----
330	----	PINERIA	----
331	3:55 "	SAN JUANITO	5:05 "
336	----	LA LAJA	----
343	4:25 "	BOCOYNA	4:36 "
	Arr: 4:45 "		Lve: 4:20 "
351		CREEL	
	Lve: 5:10 "		Arr: 4:00 "
375	6:10 "	PITORREAL	2:50 "
388	7:00 "	ING.FCO.M.TOGNO	2:15 "
	Arr: 7:22 "		Lve: 1:50 "
396		SAN RAFAEL	
	Lve: 7:32 "		
411	8:20 "	CUITECO	1:00 "
416	8:33 "	BAHUICHIVO	12:45 "
427	9:09 "	PARAJES	12:09 "
434	9:35 "	CEROCAHUI	11:45 AM
440	10:00 "	TEMORIS	11:20 "
449	10:35 "	SEPTENTRION	10:43 "
458	----	SANTO NIÑO	----
465	11:30 "	JESUS CRUZ	9:53 "
471	11:55 "	DESCANSO	9:33 "
484	12:30 PM	AGUA CALIENTE	9:08 "
492	Arr: 12:48 "		Lve 8:55 "
		LORETO	
	Lve: 12:50 "		Arr: 8:52 "
502	1:05 "	LA LAGUNA	8:34 "
511	1:22 "	ING.H.VALDEZ	8:17 "
522	1:42 "	EL FUERTE	7:57 "
533	1:59 "	NOROTES	7:40 "
	Arr: 2:30 "		Lve: 7:00 "
549		SAN BLAS	
	Lve: 2:40 "		Arr: 6:50 "
560	3:00 "	CONSTANCIA	6:30 "
	Arr: 3:30 "		Lve: 6:00 "
572		LOS MOCHIS	
	Lve: 3:45 "		Arr: 5:40 "
585	ARR: 4:10 "	TOPOLOBAMPO	LVE: 5:20 "
	WEDNESDAY AND SATURDAY		MONDAY AND THURSDAY

Effective May 15, 1962.

5.5 and **5.6.** A Chihuahua al Pacífico time table, cost of fares, and list of bridges and tunnels, from 1962.

LIST OF BRIDGES AND TUNNELS EXISTING IN THE MOUNTAIN REGION OF THE TRUNK LINE OJINAGA-TOPOLOBAMPO

TUNNELS	BRIDGES	KILOMETER POST	LENGTH IN FEET
1		345	380
2		345	411
3 "BOCOYNA"		556	1011
4 "CONTINENTAL"		563	4134
5		591	128
6		593	130
7		593	94
8		600	160
9		601	205
10		606	622
11		609	366
12		619	415
13		623	377
14		633	182
15		633	391
16		639	311
17		639	1526
	LA LAJA	640	696
18		641	1142
19		650	367
	LA MORA	651	458
20		651	807
	SEHUERAVO	652	432
21		652	568
22		653	331
23		654	705
24		655	587
25		655	322
	NOVOCHIC	656	391
26		656	673
27		656	531
28		657	499
	ROCOHUAINA	657	387
	MACHAGACHIC	661	517
	CUITECO UNO	663	210
29		663	242
30		663	404
31		664	246
32		664	221
	BAHUICHIVO	670	222
33		683	281
34		684	259
	LAS ESTRELLAS	684	103
	SANTIAGO	687	169
	RIO PLATA	688	272
35		692	2639
36		693	636
37		693	237

----->

TUNNELS	BRIDGES	KILOMETER POST	LENGHT IN FEET
38		693	1103
	VIADUCTO UNO	695	134
39		695	423
	CEROCAHUI	696	173
40 "CEROCAHUI"		967	1471
	SEPTENTRION	698	315
41		701	2549
42		702	394
	LA PAPA	702	154
43		702	453
44		702	659
45		703	2680
46		704	131
	EL OSO	704	131
47		704	459
48 "LA PERA"		706	3058
	TEMORIS	708	265
	STA. BARBARA	708	714
49		711	262
50		711	869
	MINA PLATA	711	350
51		714	325
52		714	262
	EL TIGRE	714	148
53		714	144
54		715	400
55		715	151
56		716	151
57		716	387
58		717	144
59		717	518
	EL CARNERO	717	270
60		718	886
61		718	443
62		719	640
	GALINDRO	720	338
63		724	118
64		724	131
65		725	171
66		726	197
67		726	259
68		726	167
	TINAJA	728	123
	SAN PABLO	731	180
69 "TOPOLOBAMPO"		740	820
70		741	197
	LA CASCADA	745	662
	CHINIPAS	749	955
71		754	266
72		755	361
73 "EL DESCANSO"		756	5928
	EL FUERTE	781	1638

GENERAL INFORMATION

For Mexico travel you will need proof of citizenship (poll tax receipt birth certificate, Etc.)

A valid vaccination certificate (less than Three years).

Mexico Visitors Permit, secured from Mexican - Consular Office, or at the border Mexico Immigration Office.

Mexican Customs Inspector checks baggage at the railroad depot.

Cars may be parked at lots in Presidio, or at the depot parking lot at Ojinaga.

Your personal auto may be shipped by rail to Chihuahua or Los Mochis, through arrangement with railroad freight office at Chihuahua, Chih. With notice at least 15 days before your departure.

Cost of shipping your car:

Less Car Load Rate

From Ojinaga to Los Mochis....$43.80 for each 1,000 Kgs. With a minimum charge of $65.73 US.Cy.

Less Car Load Rate

From Chihuahua to Los Mochis $34.17 for each 1,000 Kgs. With a minimum charge of $51.25US.Cy.

Automobile shipments are handled on our Freight trains.

Special tours can be arranged at Los Mochis and Topolobampo. Taxis, Taxi buses and buses are available.

Fishing, either bottom or deep sea, is superb in waters off the coast or the Gulf of California. Arrangements can be made for charter of cruisers or outboard fishing boats at your hotel or at one of the numerous docks.

Diesel-powered equipment will carry you across breath-taking scenery: desert, canyons, mountains and the seaside.

For reservations write or phone to:
F.J. SAENZ C.
GENERAL FREIGHT AND PASSENGER AGENT
P.O.Box 46
CHIHUAHUA, CHIH. MEXICO.
TELEPHONE: CHIHUAHUA 2-22-84

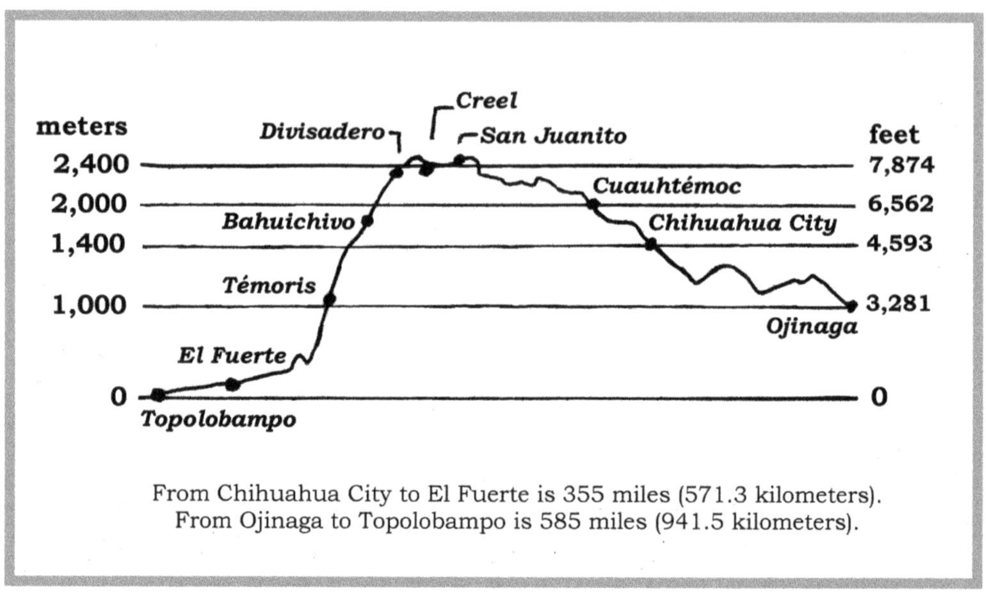

5.7. Sketch of the sierra challenge in terms of altitude.

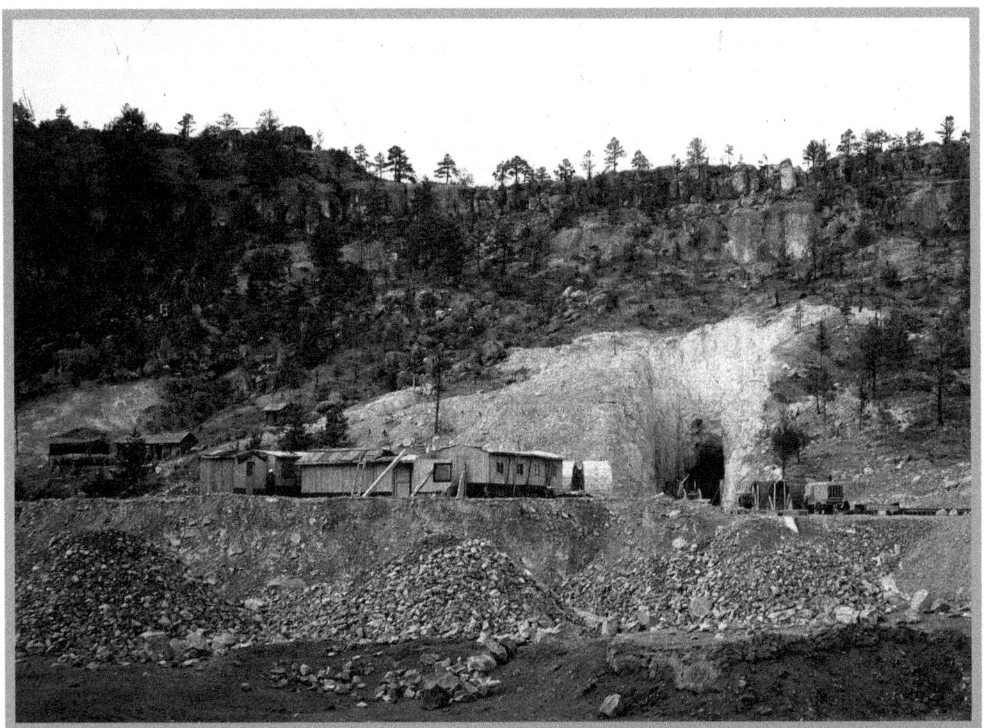

5.8. **CROSSES DIVIDE**--This tunnel will eliminate grade incline and curves. It will be 585 feet long, will cost $160,000, and is located near the Continental Divide between San Juanito and Creel, Chih. Another and longer tunnel is planned at Aguatos, west of this site (photo G.B. 1955).

After the Construction

news articles by Glenn Burgess with notes by Don Burgess

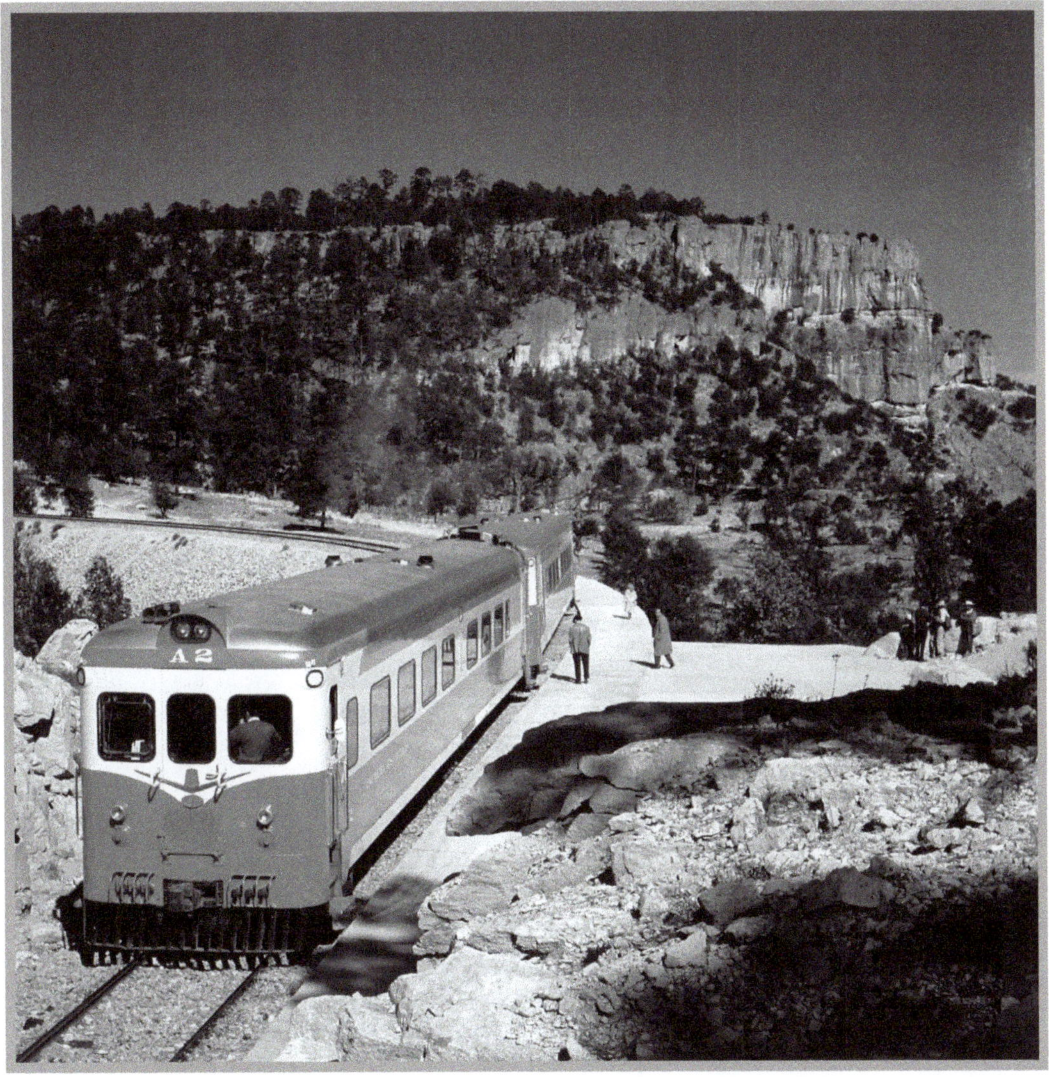

6.1. Individually powered diesel Fiat cars called Autovías were used to carry passengers for more than ten years. This photo was taken at El Divisadero: The Tarahumara man on the far right, Jesús López, and Don had just finished a 12 day hike studying the Baja Tarahumara language and were taking the train back to Creel (photo D.B. c.1963).

Accidents and Challenges to Operations

The operation of the Chihuahua al Pacífico Railroad has not been without its problems. Even on the inaugural train, there was a landslide that broke some windows in the train and one of the cars jumped the track when the engineer had to slam on the brakes in order to miss some workers (Ken Flynn, "Unspoiled Beauties of Sierra Madre Spark Mexico's Inaugural Train Trip for Newsmen," *El Paso Herald Post*, Nov. 27, 1961).

Landslides during the rainy season are a common problem. In some places, concrete tunnels had to be built so that rock slides could go over the top. There have also been some train crashes, such as the one in 1976 where 24 people were killed when a passenger train ran into a freight train that was coming out of a tunnel west of Creel. An accident that got the citizens of San Rafael excited was in the early 1990s when a railroad car came loose and started rolling backwards. It slammed into a tank car full of diesel. There was a huge explosion which sent burning diesel spraying and running down the slope and through the lumber yard. One man was killed and the office and shop of the lumber company were burned. People were jumping on trucks to get out of town and a mile away at the other end of town they were running for the mountains. Then there was a train robbery in 1998, an old west style hold-up, where a Swiss tourist was shot and killed when he tried to take pictures of the robbery. More recently, locomotive engineers have complained that on occasion the trains have been stopped and forced to carry drugs.

The manager of one of the train stations related to me an accident that happened near San Rafael. He said a flat car became disconnected from the train and started rolling downhill. A railroad worker tried to apply the brake but was unable to do so and jumped off as the car picked up speed. Coming the other direction was a work crew on a small gondola. When they saw the car coming, the crew jumped off and the flatcar crashed into the gondola. A steel digging bar shot through the air and went through the stomach of one of the workers. The manager of the train station was called and he was the one who had to pull the bar out of the man, who soon died.

The El Paso Times

- Sunday, January 27, 1963

Tourist Patterns—Even Thought Patterns—Have Had To Change

Chihuahua Al Pacífico Railroad Now is Startling Reality; Economic Impact Will Be Gradually Felt In El Paso Area

By Glenn Burgess
Times Correspondent[1]

The Chihuahua al Pacífico Railroad, the legitimate offspring of the old Kansas City, Mexico & Orient is a startling reality, yet its economic impact on the City of Chihuahua, El Paso, Presidio and Ojinaga, and other cities in Mexico and the United States will be gradual. Trade, tourist, and even thought patterns have been built up for a hundred years and these were designed to avoid the mighty Sierra Madre Occidental that has always separated the coastal sites of Sonora, Sinaloa, and Nayarit from Chihuahua, Durango and Zacatecas.

The railroad bisects the mountain range that has separated these six states. The original distance north and south without benefit of rail transportation from Nogales, Ariz., to the large city of Guadalajara, was approximately 900 miles. There was no economical way to transport the products of the fertile coastal plains of Sinaloa around Los Mochis to Chihuahua. The new scenic highway completed recently between Durango and the seaport town of Mazatlán helped, but its western terminal was too far south of Los Mochis to be of much commercial benefit to Chihuahua. The well-engineered Chihuahua al Pacífico, during its first year of operation is showing that it has opened the door to the rich agricultural region of Sinaloa, and that products from Chihuahua, especially cotton from Ciudad Juárez, Ojinaga and Ciudad

1 These later articles were written after Glenn had worked on a master's thesis on the history of the railroad at New Mexico Western College in Silver City, NM (July 1962), and include additional information.

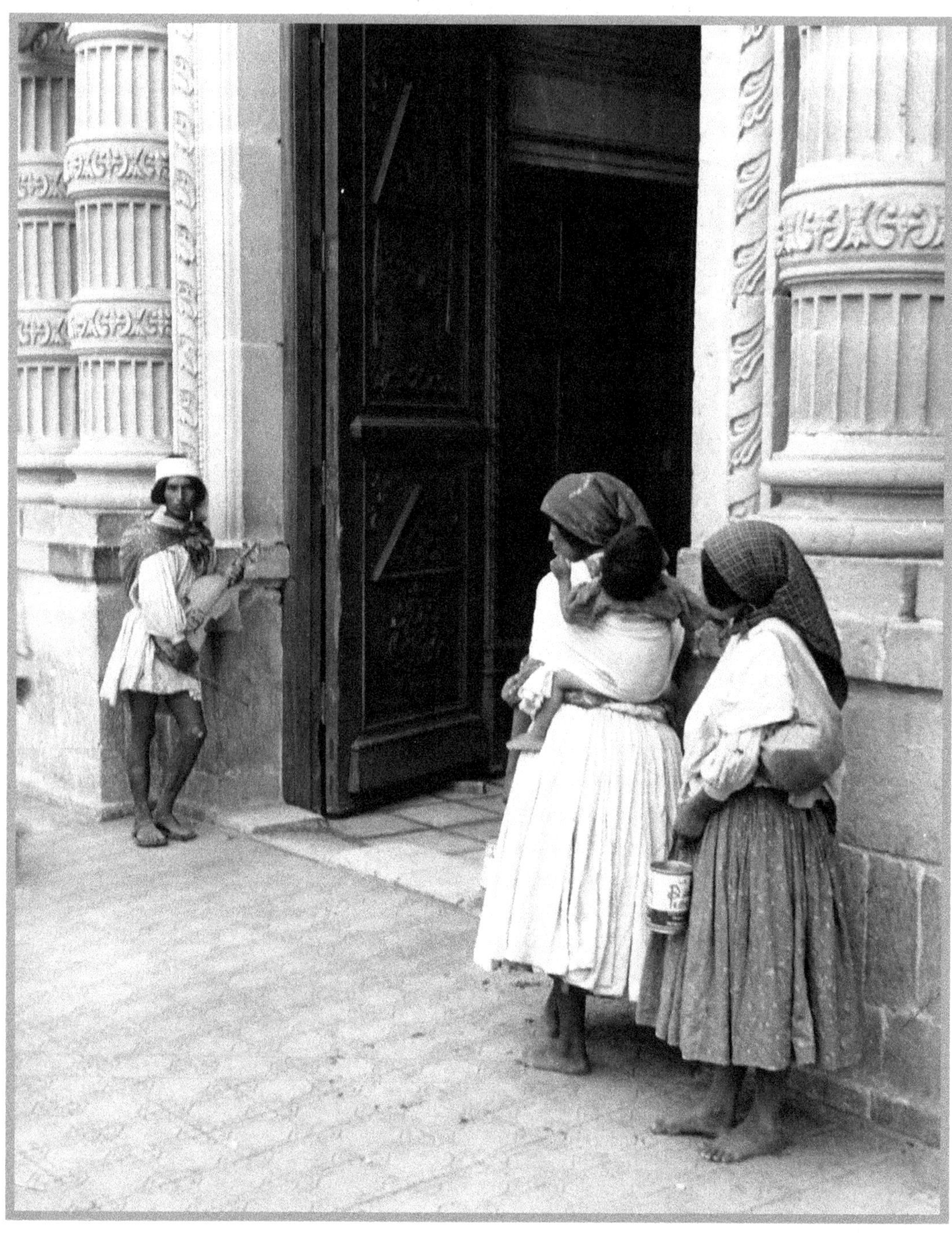

6.2. Tarahumaras outside a government building in Chihuahua City (photo G.B. 1955).

Delicias, can be transported to the seaport of Mazatlán, and when it is developed, the Port of Topolobampo 12 miles from Los Mochis. The citizens of southwestern Texas, southern New Mexico, Chihuahua, and Sinaloa are beginning to realize they have one of the most scenic rail routes in America.

Movement of commodities has been north and south for so many years, and markets for these products have been set up geographically as a result. The tourist movement has also followed the same directional pattern. Both patterns will be slow in changing.

FIRST EFFECT

The first effect on Chihuahua has been the appearance in the markets of the City of Chihuahua and of Juárez to a lesser degree, of vegetables and fruit from Sinaloa. Chihuahua now is only 18 hours from Los Mochis for movement of freight. Ciudad Juárez and El Paso are less than 30 hours away. Sugar from Los Mochis is one of the major eastbound items of freight, as is wheat and cattle feed. Cotton tops the westbound items.

After the dedication of the railroad in late November of 1961, people of both nations who had been following the progress of construction that cost the Mexican government $88,368,000 (U.S.C.), wanted to make the trip. Tourists from El Paso and southern New Mexico usually drive to Chihuahua City and spend some time there. Others come over the train from Presidio and Ojinaga. Some of these stop over in Chihuahua. Again, there is an increasing number who come from the Los Mochis end. Most of these are tourists who learn of the railroad after driving down from California and Arizona. Then, of course, there are many people from Chihuahua and Sinaloa who want to see the country and to marvel at the engineering feats created by their own Mexican engineers.

However, the monthly increase of passengers has not kept pace with the increase of freight and express.

The fact that Chihuahua City now has an east-west transcontinental railroad crossing its north-south line running from Ciudad Juárez to Mexico City is bound to be another of the stabilizing factors in the rapidly growing city of Chihuahua. The city, itself, now has around 200,000 inhabitants. This year, 1963, should be a prosperous one. According to Señor Cipriano Ortega, general

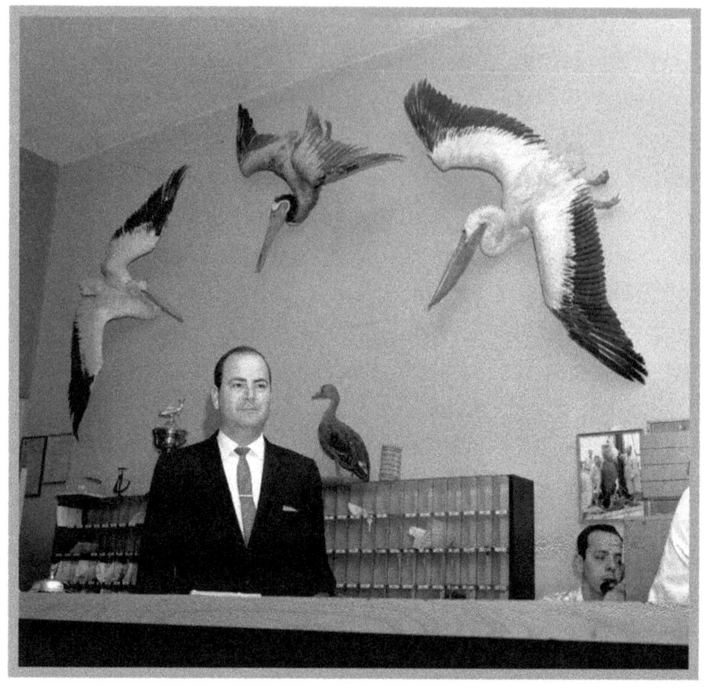

6.3. Lic. Roberto Balderrama Gómez in the Santa Anita Hotel in Los Mochis, which he opened in 1958. He later built hotels along the railroad in El Fuerte (Posada Hidalgo--1968), Cerocahui (Hotel Misión--1976), Cuiteco (Hotel Cuiteco--1977), and Areponápuchi (Hotel Mirador--1992) (photo G.B. c.1960).

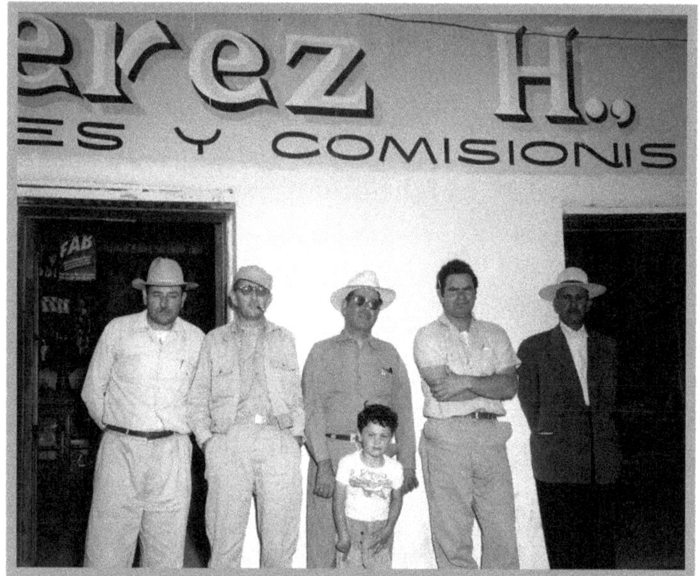

6.4. The general store Casa Pérez y H. (hermanos) S.A. in San Juanito. Man on left is Ramón Grijalva. Next to him is Jorge Togno. Second from the right is a man called Sr. Meno (identifications by Carlos Jaime). Today Abarrotes Mendoza is located here (photo G.B. 1955).

manager of the Banco Comercial Mexicano, good rains during the late summer and fall, and during December have placed the cattle industry in good shape, there seems to be an abundance of employment, and business men have confidence in the future of the city as evidenced by a competitive bank erecting a 22-story building directly across the main plaza from Señor Ortega's fine bank building. Señor Ortega believes the new railroad will help Chihuahua in many ways and that this economic assistance will "snowball" as the developments along the route come and as new trade patterns emerge. He also points to the fact that new irrigation projects in Sinaloa are being developed and new products of that area will find markets in Chihuahua, El Paso, and mid-continent United States.

Perhaps the impact has not been as great as promoter A.E. Stilwell predicted when he started the Kansas City, Mexico & Orient in 1900; or of socialist minded Albert Kimsey Owen, who arrived in Sinaloa in 1871 and decided that Topolobampo and Los Mochis was a good place to found a modern "Utopia" that would need a railroad to the U.S.; or even of the Civil War Confederate Col. James Reily, who rode down from El Paso in January 1862, to try to convince Gov. Luís Terrazas that the State of Chihuahua would be better off if it should join the Confederacy. On Jan. 26, 1862, Col. Reily wrote "Chihuahua is a rich and glorious neighbor and would improve by being under the Confederate Flag. There are no such mines in the world as are within sight of Chihuahua City. We must have Chihuahua and Sonora—by a railroad to Guaymas we render our great State of Texas the great highway of Nations."

Reily's dream of a political union with Chihuahua never came about, and the railroad missed Guaymas by 219 miles, but, according to Abel Prince, assistant general manager of the new Chihuahua al Pacífico and who has 52 years of railroading experience back of him, the line is developing the back country of the Sierra Madre, is acting as a new commercial connecting link between the coastal plains and Mexico's northern plateau region, and is living up to the expectations of Francisco Togno, who completed the engineering plans of the railroad, and of the officials of the Mexican government who raised the money for the final construction and rehabilitation of the earlier trackage.

F.J. Sáenz, director of the traffic department, also a veteran railroad man, points with pride at the development of freight and passenger traffic. In January of 1962, 3,603 tons of freight, 56,552 kilograms (62 tons)[2] of express and 3,579 passengers moved across the mountains. In November, there were 22,774 tons of freight, and 391,585 kilograms (432 tons) of express. Also 5,934 passengers crossed over in August, but the number dropped off some with the fall months.

Sáenz predicts that traffic from Sinaloa, southern Sonora, and from the Port of Topolobampo, will become a major development in the future with mid-continent United States. He points out there are two equi-distant routes from San Blas, Sinaloa, to Kansas City. San Blas is east of Los Mochis and at the junction of the north-south coastal railroad and the Chihuahua al Pacífico. By Presidio, Texas, the distance is 1655 miles. By El Paso via Santa Fe to Kansas City, the distance is practically the same. By El Paso then over the Southern Pacific, the route is 1620 miles. The distance from Kansas City via Nogales is 2,060 miles. El Paso, via Nogales is 1,111 miles by rail from San Blas. From El Paso to San Blas via Casas Grandes, Madera and La Junta, the distance is only 671 miles. According to the reasoning of Sáenz, El Paso is the first large city of the United States that stands a good chance of profiting from the completion of the Chihuahua al Pacífico Railroad. The new route still offers the Kansas City region a Pacific port 400 miles nearer than any existing U.S. Pacific ports!

Tourists will soon ride from Chihuahua City, and possibly Ojinaga, to Los Mochis and Topolobampo on a modern streamlined passenger of Italian origin.[3] From Chihuahua City to Los Mochis, it will be a daylight trip all the way and made in 12 hours instead of the average run of 16 hours at present. Part of the equipment has already arrived in Chihuahua.

Finally, projecting their belief in the future of the tourist business in the Sierra Madre, on January 15, construction work on a new hotel-motel project costing 25,000,000 pesos is scheduled to begin where the Chihuahua al Pacífico touches the Barranca del Cobre. The

2 Note that there is some discrepancy between kilograms and tons in Glenn's writings.

3 These were individually powered Fiat cars known as Autovías.

project, financed by Colonizadora del Divisadero Barrancas, S.S. a Mexican group, includes convention facilities for 500 people, apartments, long distance telephone, swimming pool, and all the trimmings. It will be the center of tours to all parts of the Sierra Madre and the 1,000 miles of Barranca Country.[4] This is just one example of the economic impact of the new railroad on the deep canyons in the Barranca del Cobre.

4 The operation of the Hotel Divisadero-Barrancas, built by Ing. Efraín Sandoval Loera, began in 1973. It is presently owned and operated by his daughter Yvonne Sandoval Almeida.

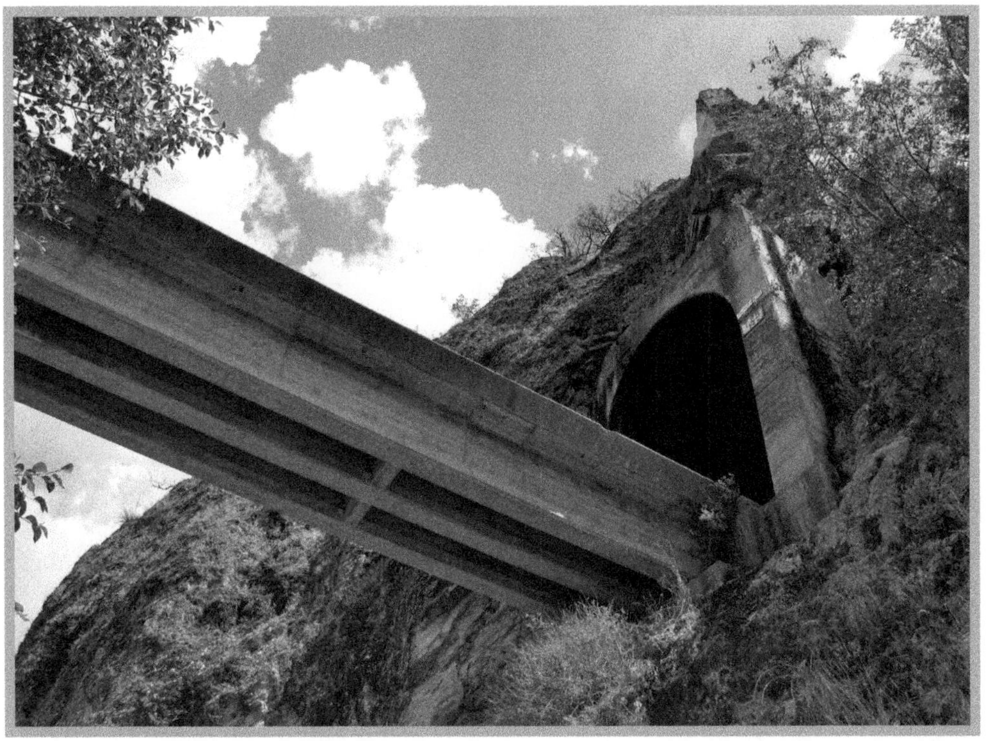

6.5. Bridge of concrete construction entering the tunnel just downstream from Cuiteco (photo D.B. 2011).

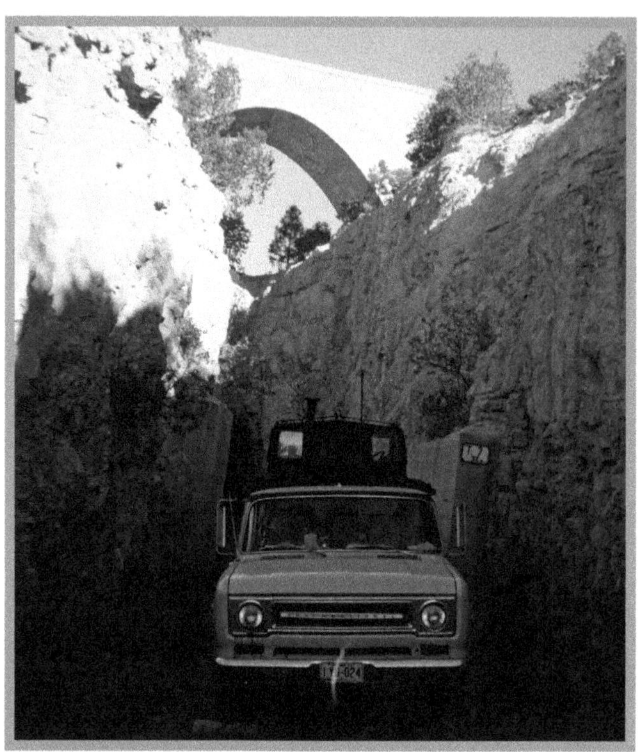

6.6. El Lazo in use: family Travel-All on flatbed next to caboose, coming out of El Lazo (photo D.B. c.1970).

Developments to Support Tourism

One tourism development was the "piggy-back" groups, sponsored by Adventure Caravans and several other companies, where tourists put their campers and RVs on flatcars for the ride across the mountains. Arriving in Los Mochis, they debarked from the train and either headed for Nogales or Mazatlán or took a ferry across to Baja California.

Back in the early 1970s, several times I put our International Travel-All on the train in order to avoid the rough roads between La Junta and San Rafael. The first time I did this, I assumed that the railroad would have chains and whatever was needed to tie the vehicle down. I was wrong, and I had nothing with me. I searched around and could only come up with a few pieces of wire. The vehicle was shifting around considerable as the train stopped and started and two railroad workers that came by shook their heads and said they didn't think this one was going to stay on. At one point I jumped off and grabbed two railroad ties to put in the front and back, but they were not tied or nailed down either. I kept my foot on the break the whole distance and we made it. I had my family and almost everything I owned in that Travel-All.

Another development was the Sierra Madre Express that had five classic rail cars from the 1940s and 1950s, including a domed dining and observation car which ran tours from Nogales, Mexico into the Tarahumara area. It is too bad that Arthur Stilwell was never able to take his private railway car, complete with organ, across the mountains. I can imagine him gathering the train crews for observance of the Sabbath (See Kerr, p.72).

The El Paso Times

– Sunday February 3, 1963

Mountains, Tunnels, Canyons Impressive

Spectacular Scenery Yields Impact To Chihuahua Al Pacífico Train Ride

By Glenn Burgess
Times Correspondent

Within easy reach of the people of southwestern United States is one of the most scenic and scientifically engineered railroads in America. The Chihuahua al Pacífico Railroad does not climb more than 8,500 feet above the sea, but the scenic grandeur of the Barranca Country between Chihuahua City and the Gulf of California is certainly different than that encountered by any other railroad in America. To many passengers, the route is even more scenic and more interesting.

Completed as an economic necessity, the railroad breaks through the canyon-gashed barrier formed by the Sierra Madre that stretches 900 miles from Douglas, Ariz. to Mexico's second largest city, Guadalajara. This timbered region no longer separates the rich agricultural states of Sonora, Sinaloa and Nayarit from Chihuahua, Durango and Zacatecas. The final 156 miles, though costing $86 million, is rich in mineral and timber resources, and a potential tourist mecca.

The Chihuahua al Pacífico, begun in 1900 as the Kansas City, Mexico & Orient, not only ties northern Mexico together, but it fulfills the original dream of early promoters who saw a Pacific port 400 miles nearer Kansas City than an existing Pacific port in the United States and who predicted that mid-continent United States would have need for the products of the 'bread basket' of Mexico. While the railroad is accomplishing these feats, it will attract many new tourist dollars to help balance Mexico's national budget.

The first economic impact felt seems to be the transportation of goods and people between the states of Chihuahua and

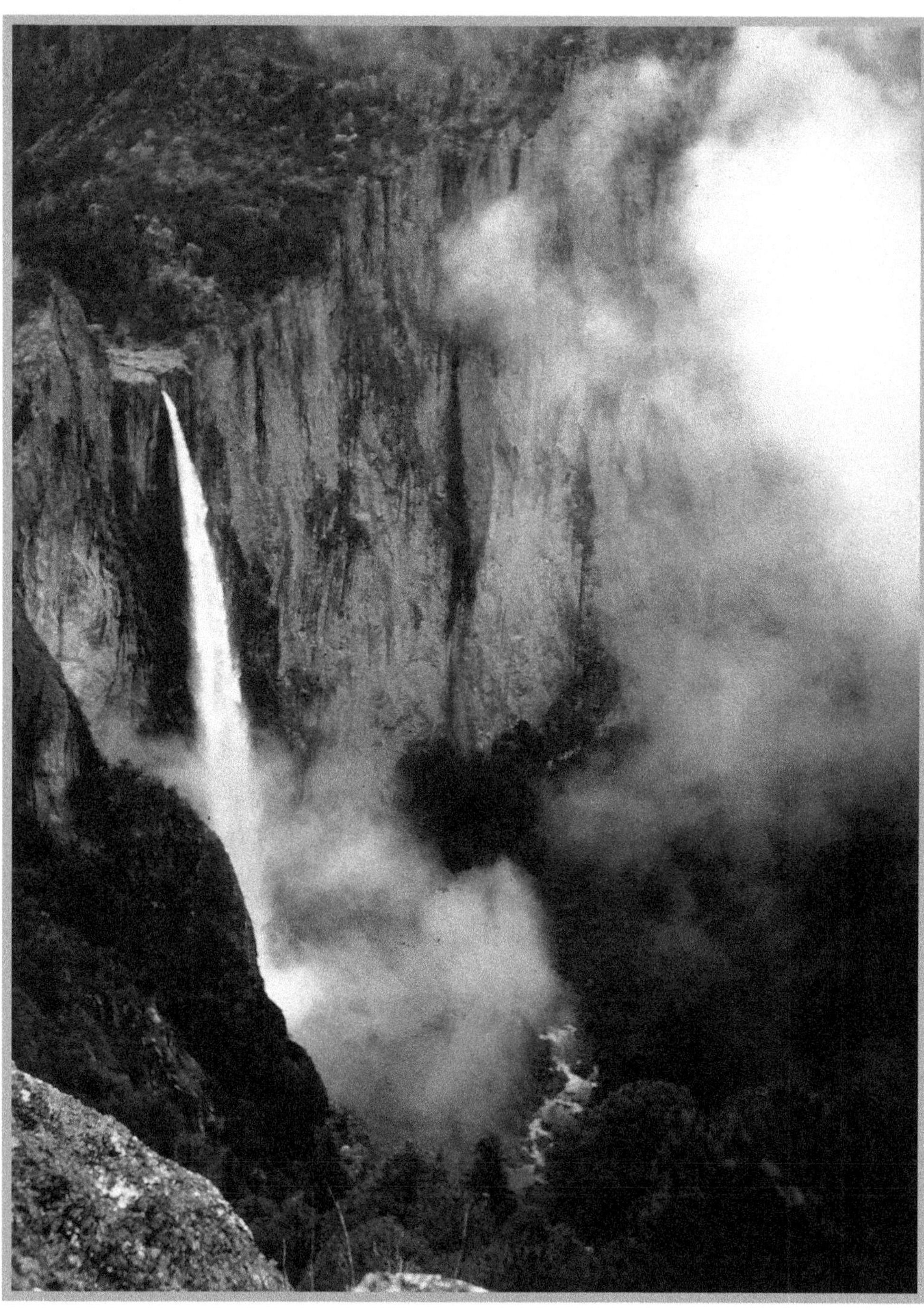

6.7. Basaseachi Waterfall, 807 feet (246 meters) high (photo D.B. c.1980). Just before the water goes over the edge, the water has made a deep cut into the rock. A few feet back from the edge of the falls, a large rock lies across the cut. Once, when Glenn was crossing over this rock, he slipped and fell, breaking his ankle. He was within inches of falling into the rushing water, whose strong current would have swept him over the fall.

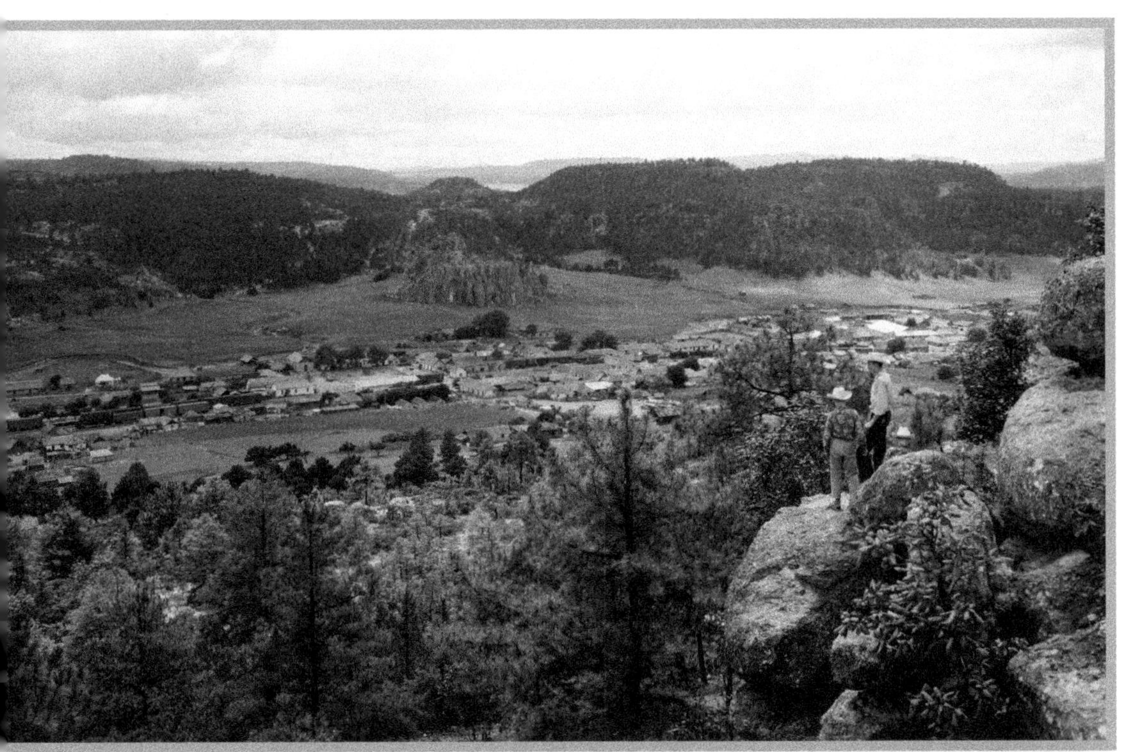

6.8. From 1907 to 1961, Creel was the western end of the railroad (photo G.B. 1960).

Sinaloa. The timber and mineral resources between Creel and El Fuerte, the terminals of the last section of 156 miles finished in 1961, are still latent. However, there is little question that sawmills and mines will develop soon due to the now available rail transportation. If properly promoted and developed, the tourist potential is just about unlimited. The thousand miles of canyons comparable to the Grand Canyon, the many mountain torrents, high waterfalls up to 900 feet,[1] the variety of climates from semi-tropical to high mountain zones, and the Tarahumara Indians will attract both the Mexican and the United States tourist. The proposed 25 million peso hotel to be built at the point where the Chihuahua al Pacífico touches the fantastic

1 The actual height of Basaseachi Falls, according to Dr. Schmidt, is 807 feet (246 meters). He, along with others, measured the falls using a steel wire that would not stretch, and the latest survey equipment. Once, just above the falls, when Glenn was crossing over a boulder that spans where the water has cut down into the rock, he slipped and fell, coming within inches of falling into the water, which would have swished him over the falls.

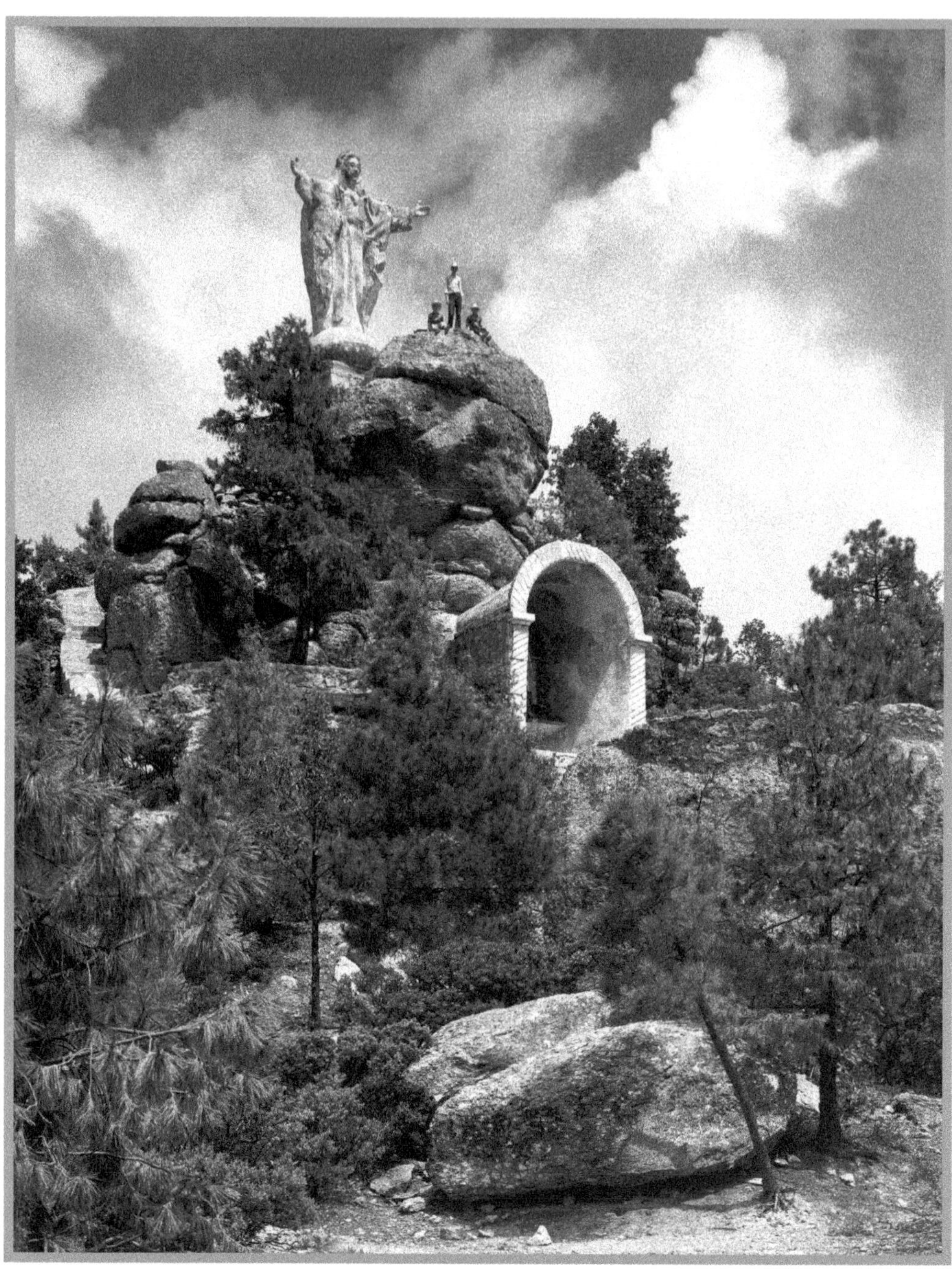

6.9. FACING DIVIDE—This shrine with its large statue of Christ, called Cristo Rey, faces the Continental Divide and overlooks the Creel Valley (photo G.B. 1960). [This statue was later replaced when lightening broke off one arm.]

Barranca del Cobre, is the best tangible evidence that the tourist business will move in.

BOOM TOWNS SUFFER

At present, the little towns along the Creel-El Fuerte sector are suffering from the effects of dropping back to normal after they had become construction boom towns. Much of the $88 million construction costs went to labor. Sixty-seven Mexican engineers and 9,300 workmen built the railroad during the construction period that began in 1940 and increased in intensity until the final years of 1959-61 when two construction companies were employing at one time more than 6,000 men and during which time the government spent $57 million. Creel, just over the Continental Divide, a little town of 1,200 people, had a floating population of 3,000! Small wonder that it looks a little dreary now. The same thing in proportion has happened to the other places. These villages will have to wait for the development of connecting roads and of local lumbering, mining, agricultural and tourist projects.

The states of Durango, Sinaloa and Chihuahua have 46.4 per cent of the standing merchantable timber of Mexico, and the Chihuahua mountains traversed by the new line have 51 per cent of the pine, fir, and spruce reserves of the nation. The production of forest products in Chihuahua during 1960 amounted to $12 million and the mountain timber belt extends many miles southwest of the former railway terminal of Creel.

Most of the mountain country along the new railroad is or can be grazed by cattle. The new line will save many tortuous miles of cattle driving and will open up new areas where it was impossible to sell local cattle. Sonora, Sinaloa and Chihuahua now produce 13.2 per cent of the cattle of Mexico. Annual production of the three states amounts to $68 million.

The potential mineral production of the area traversed by the line is still an unknown quantity. However, the stories of fabulous operations closed down during the revolutionary period at such places as Batopilas, Chínipas, Urique, La Bufa, and many others, cannot all be wild rumor.[2] Known minerals in the region are lead, zinc, mercury, and especially copper. The latter abounds in lead and zinc deposits that are

2 One of Chihuahua's largest gold mines, El Sauzal, is now located not far from where the Urique River meets the San Ignacio River.

said to be near the new line in Chihuahua. Present production of minerals in Chihuahua, Sonora, and Sinaloa amount to $90 million annually. The 156 miles of new railroad runs through the heart of this fabulous mining region.

SEVERAL FUNCTIONS

The railroad will perform several functions in the development of tourist business. In the first place, the scenic beauty of the canyons, mountains, rivers and waterfalls will attract many people. Some will just want to ride the railroad and view it from the train windows. This alone will be worth the expense, time and effort of making the trip. The Barranca del Cobre and the Septentrión Canyon are the highlights. Others will want to stop over in the region to explore the barrancas, or to enjoy the summer or winter climate of the scenic region. Hunting and fishing and short plane trips over the barrancas will be added attractions. Others will be interested in meeting the Tarahumara Indians and a few will have the stamina to visit these people in their homes in the barrancas. Again, that group of people who have taken up railroads as their hobby will find the 156-mile sector between Creel and El Fuerte will have just about anything in modern railroad engineering imaginable. Finally, the new railroad is becoming a new route for deep sea fishermen from Texas and New Mexico.

Tourists boarding the line at Ojinaga across the Rio Grande from Presidio pass through semiarid country traversed by mountain ridges of limestone. This 165-mile sector to Chihuahua City is very typical of the Mexican central plateau region and is a part of the Chihuahua Desert. From Chihuahua City, the line climbs rapidly and passes through country similar to that of Silver City, N.M., or Alpine and Marfa [Texas]. Sixty-four miles west of Chihuahua City it crosses a ridge 7,800 feet above sea-level, then drops down into the wide valley that contains La Celulosa de Chihuahua, a mill making paper from pine; and Cuauhtémoc, the largest town between Chihuahua City and the State of Sinaloa. The thriving dry-land farming colonies of the Mennonites have built Cuauhtémoc.[3]

Soon after leaving Cuauhtémoc, the train goes over the Continental Divide, a spot that is not too impressive as far as

3 Today much of the farming is by irrigation.

scenery goes.[4] Then it passes through La Junta where another segment of the Chihuahua al Pacífico turns north through Madera and Casas Grandes on its way to Ciudad Juárez. A few miles west of La Junta is encountered the old mining town of Miñaca, once about as turbulent as Cripple Creek, but now almost deserted.[5] Just 25 miles west of Miñaca and 145 miles from Chihuahua City, the pine belt of the Sierra Madre begins. A few miles farther on, the railroad again crosses the Continental Divide to the headquarters of the Rio Conchas that joins Presidio. This time the Divide is more interesting and the lumbering center of San Juanito is picturesque.

LONG TUNNEL

Before reaching Creel, the railroad goes again to the Pacific side of the Continental Divide, but this time it goes under it, through a tunnel that is 4,134 feet long. Two miles farther, the train stops in Creel, that from 1907 until 1961, was the western terminal of the Chihuahua-Creel segment of the rail project. Eighteen miles farther west, the line reaches its highest elevation—8,500 feet. From this point back to Miñaca, the country has been low mountainous terrain with beautiful forests of Ponderosa pine, oak, and even some aspen and fir. There are a few running streams, some caves in the volcanic tuft where Tarahumaras live in the summer, and small corn patches where in the summer, oxen are still used and where in the winter deep snows may cover the ground.

From the high point west of Creel, things begin to happen rapidly to the scenery. Deep canyons start opening on either side of the railroad, tunnels become more frequent, the train twists and turns to get around the mountains. Once it makes a complete loop and crosses over its own track on a stone bridge where the lower section enters a tunnel. In the same area the train goes through a hand-dug cut that is much over a hundred feet deep. From this point the railroad drops 40 feet per mile until it crosses the Rio Fuerte on a bridge that is 1,638 feet long and 146 feet above the mighty river that furnishes enough water to irrigate

4 This place, called Mal Paso, was the scene of one of the first battles of the Mexican Revolution of 1910. Pancho Villa's General Pascual Orozco positioned his men on a ridge overlooking the railroad, blocked the oncoming train which was full of Federal soldiers, and proceeded to let loose a deadly barrage of rifle fire.

5 Glenn must have meant the mining community of Cusihuiriachi.

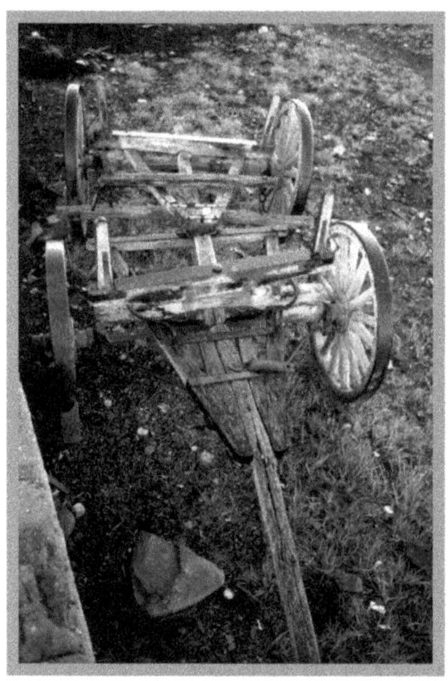

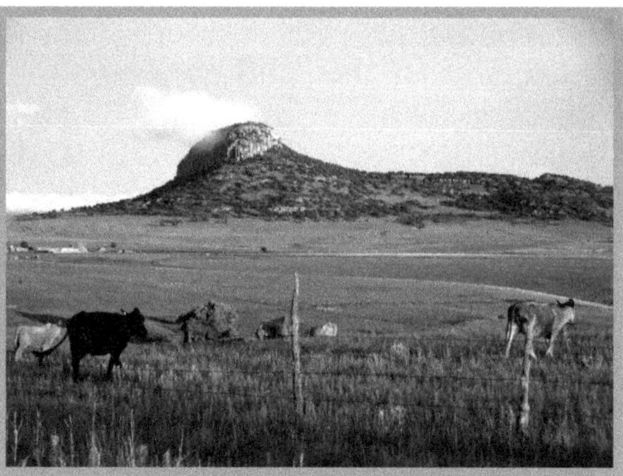

6.10. wagon frame at Miñaca (photo D.B. c.1970); **6.11.** Cerro de Miñaca (photo D.B. 2012).

Miñaca

The village of Miñaca, located on the far side of the mountain in the photo, was developed as a railroad town in the early 1900s. It was one of the main camps for the Pershing expedition looking for Pancho Villa (See *Chasing Villa* by Col. Frank Tompkins, Chapter 28). The frame of a freight wagon from that expedition is seen in the above photo. One of the wheel hubs says Kentucky Wagon Co., Louisville, Ky. Another says International Harvester Co., Chicago, Ill. And a third has Winnona, Minn. on it. The wagon is now in the Pancho Villa State Park in Columbus, NM.

At one time, Miñaca was to be the junction with the Noroeste Railroad, but politics, say the locals, put it at La Junta. The name Miñaca perhaps comes from the Tarahumara word for Mountain Lion "mawiyá". The nearby mountain looks like a lion that is lying down. Another possibility is that the word is made up of locative expressions such as 'mi "there" and na "place". Tarahumaras often use locative expressions in their place names and the Cerro de Miñaca, or Cerro de la Campana as some call it, is located in the middle of a huge flat area and can be seen from long distances. When these indigenous names are very old, the pronunciation sometimes changes and the original name and meaning are difficult to identify.

700,000 acres of fertile coastal land in a semi-tropical climate around Los Mochis.

Also, from the high point, the passenger, if it is daylight, can begin to glimpse the spectacular vistas of the Barranca del Cobre and before he is ready for the spectacle, the mighty canyon itself appears. The train, on its return trip, stops at El Divisadero just a hundred yards from the brink of the mile-deep gorge. Passengers are allowed to rush over to the viewing platforms for a brief glimpse of the colorful canyon made famous by the late José Gándara [of the Pemex Tourist Department] of El Paso and Mexico. Before the passenger realizes just what he is being permitted to view, the conductor is rushing him back to the train. Later, the tourist may return to enjoy the views, and if he is tough enough, go down into the canyon where aspen and fir are encountered on top and oranges, bananas and sugar cane grown by Tarahumaras at the bottom. Below the rim, and visible to the spectators, are corn fields on land so steep the Texas farmer would be afraid of falling out of them.[6]

DEEP CANYONS

Past the Barranca del Cobre the line continues for a short time on the top of the volcanic plateau that hides over a 1,000 miles of deep canyons. Then it begins to drop rapidly to the headwaters of the Rio Septentrión and past Cuiteco with its high quality delicious apples formerly packed out by mule and burro.

In a short distance the train enters the section of the Septentrión Canyon that proved to be the main engineering and construction obstacle to completing the line. It was the last sector finished.

Excluding tunnels and bridges, the grades and drainage alone cost $3,000 per mile. It was here that much of the $57 million appropriated by the Mexican Government in 1959-61 was spent. It was this impenetrable 27 miles of deep canyon that caused the writer to have to travel 1,200 miles in 1955 just to get to the lower end.

However, the passenger on the comfortable train knows little of this background. He sees

6 Tarahumaras tell me that when they are plowing on these steep canyon sides with two oxen strapped to a yoke, if one of them slips and falls, it could break its neck. It is interesting to note that the Tarahumara language has an elaborate system of words and endings that express the ideas of uphill, downhill, upstream, downstream, how steep the terrain is, which side of a hill something is on, etc. See Burgess, "Western Tarahumara".

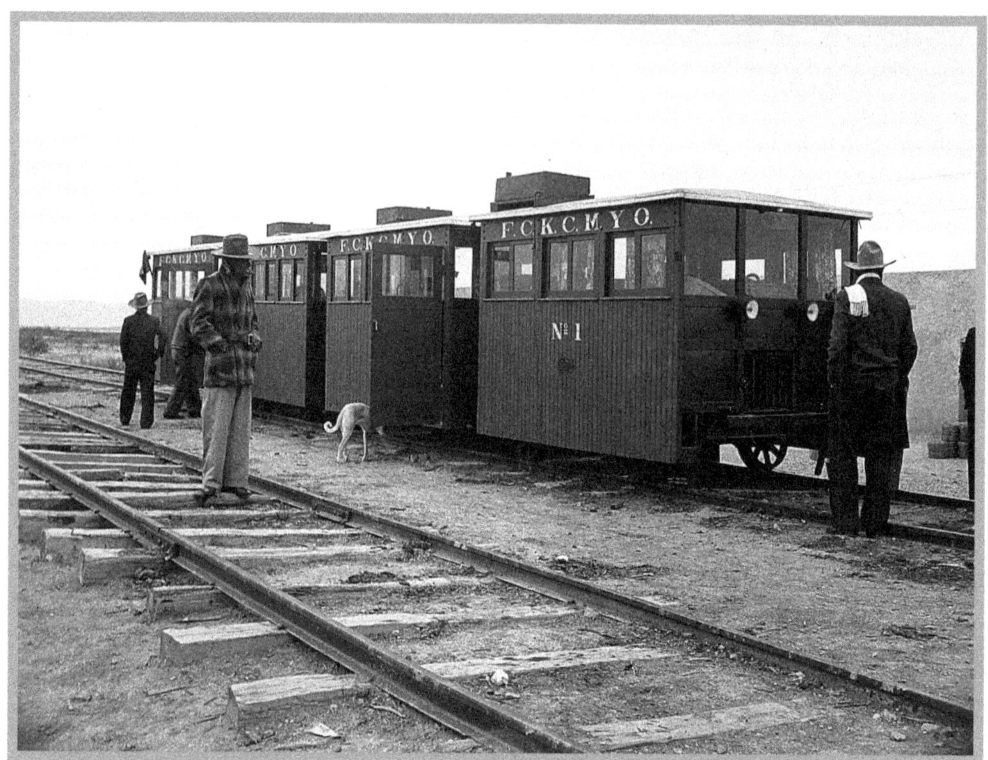

6.12. **EARLY TRAIN—**A wartime emergency train of 1942 [near Ojinaga] wasn't so modern. Powered by a tractor motor, it had four cars that were without springs. The train ran about 30 miles per hour—much faster than the regular mixed train (photo G.B.).

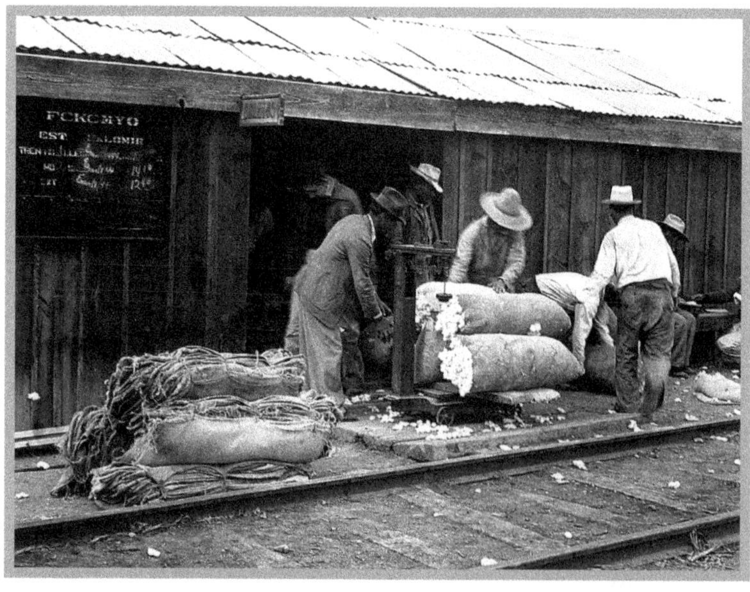

6.13. Cotton being weighed at Falomir, between Ojinaga and Chihuahua City (photo G.B. 1942).

only the myriads of mountain towers, forest covered slopes, rushing streams and deep lateral canyons where he could take a "thousand" pictures and never duplicate the scene. Remnants of construction camps cling to the sides of the canyon as the diesel powered train eases smoothly by.[7] Tunnels by the dozen begin to obstruct the view.

The most fantastic part of the entire westbound trip is the nine-mile sector beginning at Cerocahui[8] and ending at the lower end of the Chicural. At Cerocahui the train turns left into a lateral canyon where it passes through a circular tunnel 1,471 feet long before the line returns to the main canyon. Just before reaching the Chicural Basin, the train goes through another long tunnel with 2,543 feet of darkness, and coming out into brilliant sunshine, passengers see the Septentrión River many feet below them. They have just come under Témoris Falls. To traverse a linear distance of 1.6 miles, the train must travel 4.4 miles that takes it through another "U" tunnel, La Pera (The Pear), 3,058 feet long. In this nine-mile sector, the train travels through nine tunnels and over seven bridges.

Passengers have a short time to jump off the train at Témoris to take pictures of the canyon walls and the tortuous route of the railroad as it seeks not to exceed the maximum grade of 2.5 per cent. Definitely, the Chicural and the Barranca del Cobre view are the highlights of the trip across.

MAJOR BRIDGES

Another sector of three miles downstream from the Chicural has 10 tunnels and five more major bridges. The rest of the Septentrión Canyon is about the same story. The train finally breaks out into the open and literally hurtles the southbound Rio Chínipas over the highest bridge of the route. The rails are just 300 feet above the water. After the high bridge the passengers get wonderful views of the 3,000 foot western drop of the Sierra Madre. Still four miles farther along is encountered the longest tunnel, El

7 One of the former construction workers told me that when he now rides the train, he is filled with nostalgia as he looks out the windows where the construction camps used to be—Areponápuchi, La Laja, Guasachique, Los Táscates, Oribo, Parajes, Bawina, Bahuichibo, Tacuina, Piedra de Lumbre, Santa Bárbara....

8 This is a stop just below Bahuichibo. The Tarahumara word for Cerocahui is Serógachi, which I believe refers to a jagged ridge that looks like the leg of a grasshopper, located near the town of Cerocahui. The Tarahumara place name Ba'wichibo refers to a misty flat place.

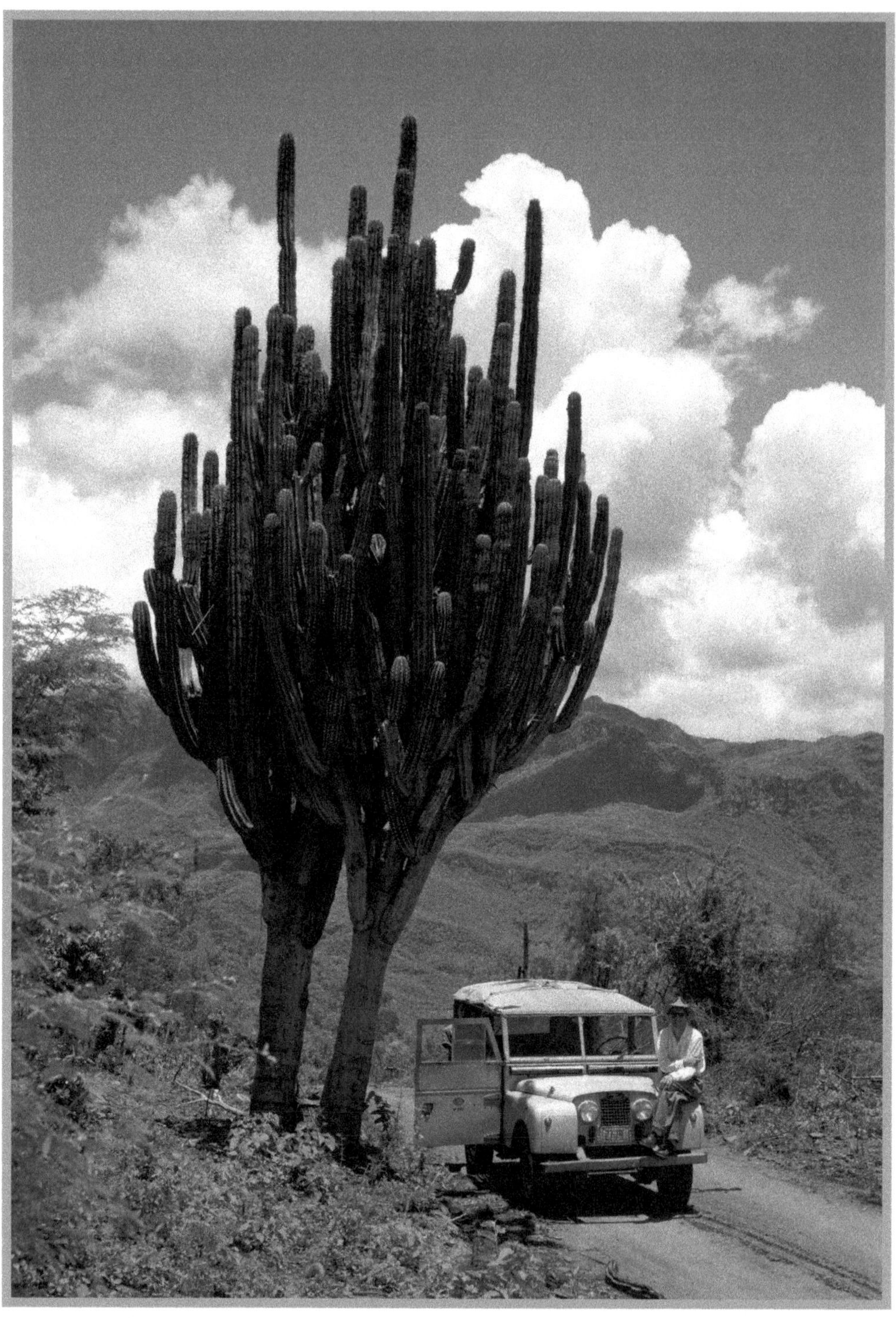

6.14. Jorge Togno posing by an Organ Pipe (Columnar) Cactus in Sinaloa (photo G.B. 1955).

Descanso, 5,928 feet [Official length is 6024 feet], or over a mile long. Before the rails were laid, it was used as a highway. Drivers entering the tunnel had to guess if the pinpoint of light was daylight or automobile headlights!

Soon the low foothills covered with the tropical variety of the giant organ pipe (columnar) cactus disappear as the railroad crosses the Rio Fuerte 12 miles above the earthen Miguel Hidalgo Dam. This dam stores most of the flood waters of the tame looking stream that becomes a roaring flood in July, August and September, the rainy season of the Sierra Madre.

Lush flat irrigated farm lands are a release after being boxed in by the towering walls of the Septentrión Canyon, but after some very interesting days at Los Mochis, and deep sea fishing out of Topolobampo, the return trip of 418 miles to Chihuahua City is still extremely interesting.

By this time the passenger has had time to become interested in the history of the line, the engineering problems involved, and the stature of the task accomplished by Mexico. The evening shadows produce a new perspective and the Sierra Madre appears even more beautiful than on the trip down. It is doubtful if any passenger fails to say: "This is the most fantastic trip I have ever taken on a train. Words cannot describe the scenery. The railroad, itself, is truly 'a work of the Romans'!"

Laughing about Speed (or lack thereof) in the Sierra

The challenging geography of the Sierra meant that engineers had to use frequent switchbacks. Distances that were short if flown directly were multiplied when negotiated by rail or road. In addition, the curves and gradients required a slow speed. In some locations, it was faster for a person raised in mountain climbing or even for a visitor to the Sierra to walk the distance.

This was especially true of the early days of the train. In 1930, when the anthropologist Robert Zingg was traveling from Chihuahua City to Creel on the train, he wrote:

> The great horseshoe curves taken by the road in climbing the sierra are so wide and the gradient so steep that in one place I dropped off the train at the beginning of a curve and walked the intervening distance. I caught the train on its way back, with time enough to light a cigarette. We had gained about fifty feet in altitude perhaps, but were virtually no nearer our destination (p.4).

A story told by Erle Stanley Gardner in 1954 (*Neighborhood Frontiers*, pp.225-226) also elicits a laugh. Gardner was famous for his mystery novels, but this book tells of his travels in Mexico. He wrote:

> Creel is the end of a railroad line of sorts which runs with a great deal of whistle blowing and puffing. Derailments are frequent, but the speed is such that no one could possibly get hurt.
>
> In fact, there was a delightful story about the railroad that the Mexicans related with gusto.
>
> It seemed that the afternoon train was particularly slow, wandering along and stopping, until finally it came to a dead stop with the engineer tooting frantically on the whistle.
>
> An American passenger, who could speak some Spanish, called the conductor and demanded to know what was the matter.
>
> The conductor, with much smiling and shrugging of the shoulders, explained that there was a cow grazing along the track. The engineer was doing his best to get the cow out of the way. Soon they would proceed. "Have no fear, señor. The train will go, and soon. But in the meantime there is nothing to be done, except speak to the cow with the whistle."
>
> So the disgruntled American settled back and finally dozed off to sleep.
>
> He was awakened with a jolt as the train started and the conductor

came back to tell him, "Yes, indeed, it is as I said. No? The train starts. Now we are going along at good speed. No?"

The American had long since ceased to look at his watch. He nodded wearily and settled back to wait.

The train ran along for another forty-five minutes, then abruptly ground to a stop. Once more the engineer started tooting the whistle. The conductor anticipated the American's complaint. "Señor," he said, "I will see, I will see."

He dashed to the front of the train and then came back smiling affably, shrugging his shoulders.

"La vaca," the conductor said, his face wreathed in smiles.

"Good God," the American said, "not another cow!"

"No, no, no," the conductor expostulated. "Not another cow! It is the same one, señor!"

Little wonder the train was sometimes referred to as "El Cansado" (the tired one) as opposed to "El Kansas".

Although modern transportation technology might seem an improvement for getting places, the Sierra's geography often wins the battle. Perhaps many people, faced with traversing the mountains and canyons, are like the Tarahumara man in the 1965 movie called *Tarahumara*, staring López Tarso. The man was walking along when a government pick-up truck stopped and the driver offered him a ride. "No thank you," the Tarahumara replied. "I am in a hurry."

Sometimes it is quicker to walk.

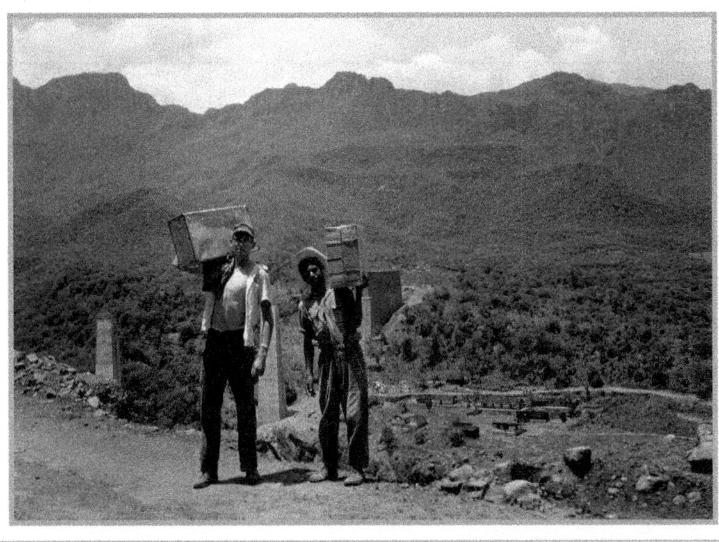

6.15. With the Chínipas bridge in the background, Don and colleague carry their suitcases, leaving a truck mired in the mud behind them (photographer unknown 1959).

The Lives of Two Railroad Workers

interviews by Don Burgess

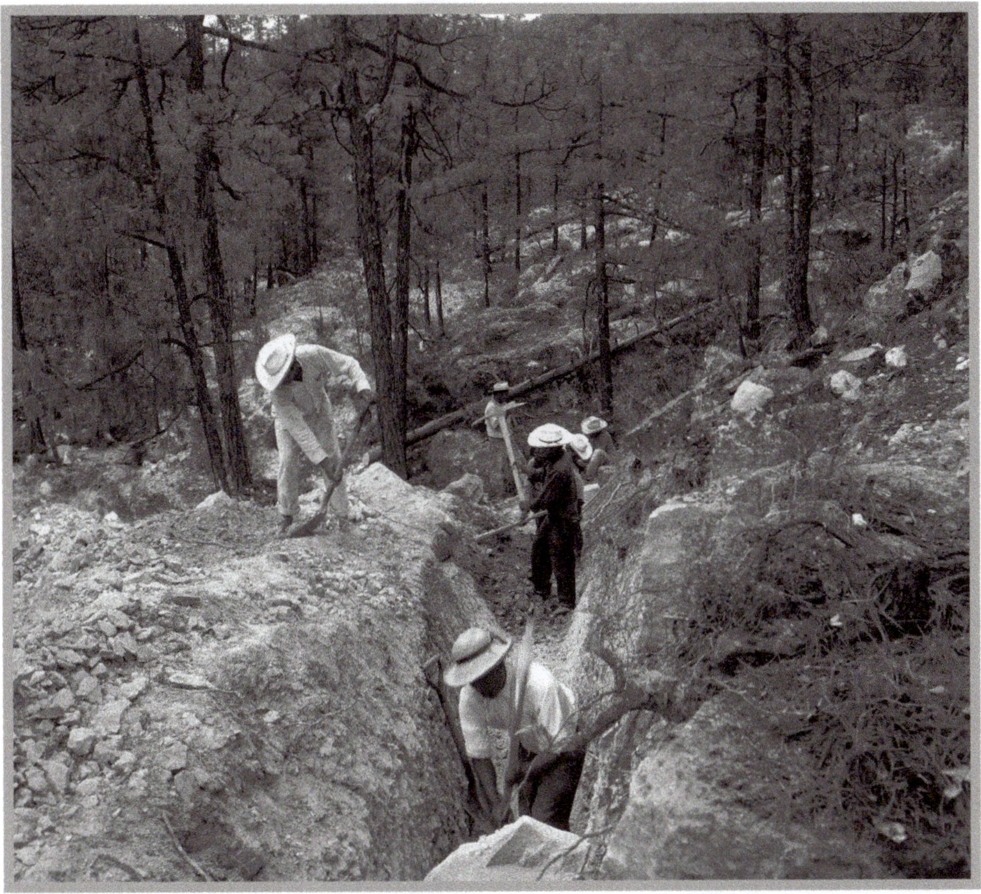

7.1. Tarahumaras working on the construction (photo G.B. 1955).

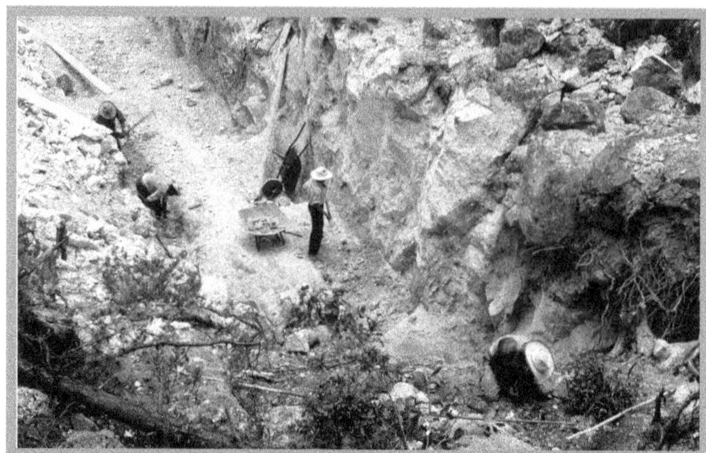

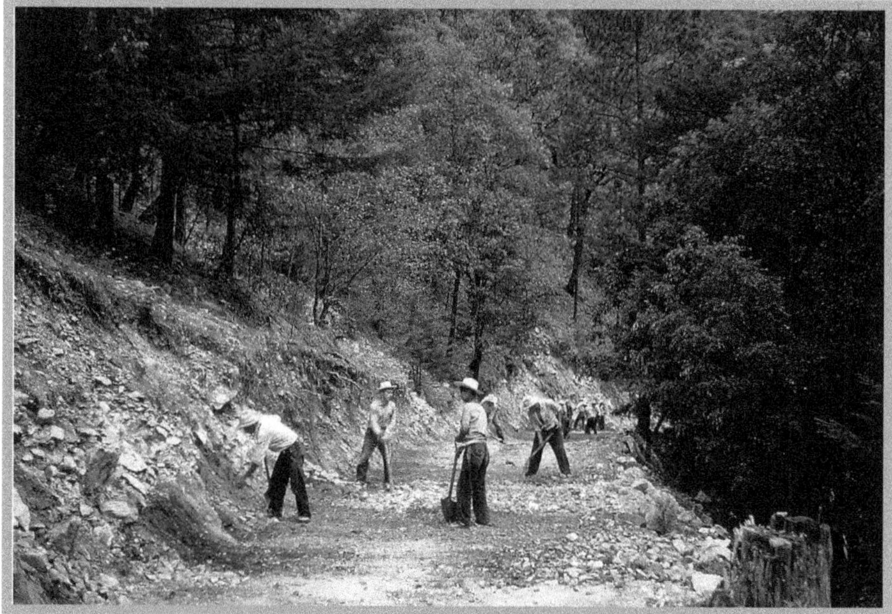

7.2 and **7.3.** Manual Labor (photos G.B. 1955).

Payment for Services Rendered

According to historian Carlos Jaime Morales, his father (Guadalupe Jaime Mora), in 1944, worked in the El Lazo-Pitorreal section as "Keeper of Time" (*Tomador de Tiempo*). His job was to keep track of the number of wheelbarrow loads of material each person produced in a day, usually around a hundred, as well as keep track of the number of dynamite holes drilled, so the workers could be paid. He had to be an honest man who could not be bribed by the workers. These particular dynamite holes, by the way, were drilled by hand. They did not have compressed air drills.

Candelario López

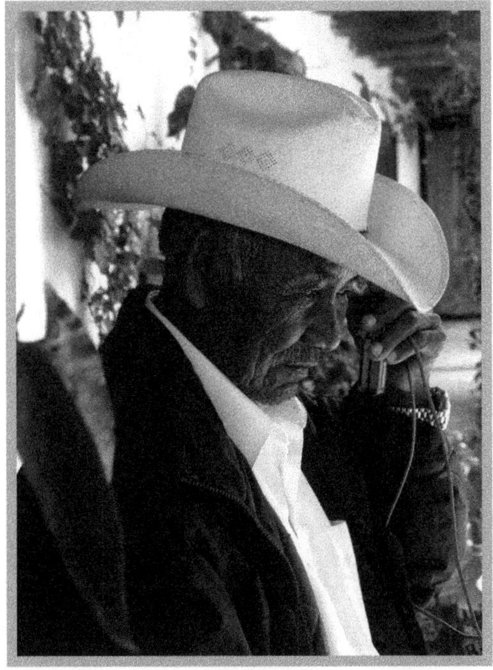

7.4. Railroad worker, Candelario López, listening to a solar operated player with the New Testament in his Baja Tarahumara language (photo D.B. 2004).

Candelario was a Tarahumara man from the Cuiteco area. Born in 1930, he died in 2011.

At 15, he began working on the railroad starting at Creel, where the line ended, helping with crews surveying where the roadbed and tunnels would be put. He worked with Ing. Lara.

For a time, he was cutting railroad ties. The most he could ever do was 25 in a day (most workers did 10-15 a day), squaring off the eight foot logs which had been cut with a cross-cut saw. He made 75 centavos per tie. It was not easy to cut the ties square and an older relative who was squaring ties sometimes would cry, not being able to get them just right.

When he was working surveying the line, the engineers would always send him to the difficult, dangerous places. In some places he would have to go down by rope. He was always amazed how the surveyors could always make a tunnel come out just at the right place, or meet perfectly in the middle of a tunnel. At the bridge sites, he would help drill core-drillings when the engineers wanted to know how stable the land was below where they would put a bridge. Usually they would drill about 25 meters. They would almost always be in some kind of rock, usually soft ash, but sometimes hard rock. At the Témoris bridge, they drilled 100 meters, but never hit rock, so they had to make extra big pilings. They have never had any trouble there, but Candelario said that at other bridges, there had been some movement, like joints spreading apart just enough to give the engineers worry.

Getting the dirt packed just right where the rails were to be laid was quite a job. After the worker had finished, one of the "jefes" might come along and throw a bucket of water on the dirt to see if it ran off like it should, or if it soaked in.

The job was a good job, said Candelario--they always paid every two weeks. When he was working in the tunnels helping to drill for blasting, it was often very wet, with water dripping from the ceiling. Once there was a cave-in while they were drilling, and a rock hit one of the workers. The one entrance to the tunnel was completely cut off, but they were able to get out the other way through a small passage.

He worked until the line was about finished (1961), then farmed for a while in Cuiteco, then went back to work for the railroad, where he worked until he retired. He was invited to the inauguration at Témoris, but did not go.

To get part of his retirement, he had to go to Chihuahua City once a month. The rest he got in San Rafael, where he was living in a house right next to the railroad line. He had a number of well-kept peach and apple and other trees in his yard. He lived there with his second wife, a Mexican lady. The first, a Tarahumara, did not want to move with him to the different places he worked.

A few years before he died, he left his second wife and was living in Creel with one of his children. He would occasionally go to Cuiteco to see his first wife.

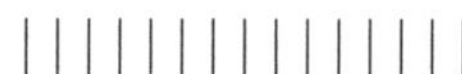

Ing. Miguel Leal remembers once when Candelario was in charge of thirty workers cleaning out rock in a cut, he suddenly felt tiny rocks hitting him in the face. Candelario realized they were shooting out from the sides of the cut and were caused by tremendous pressure from above and that the wall was about to give way. He shouted to the workers to get out of the cut, just seconds before the whole wall caved in.

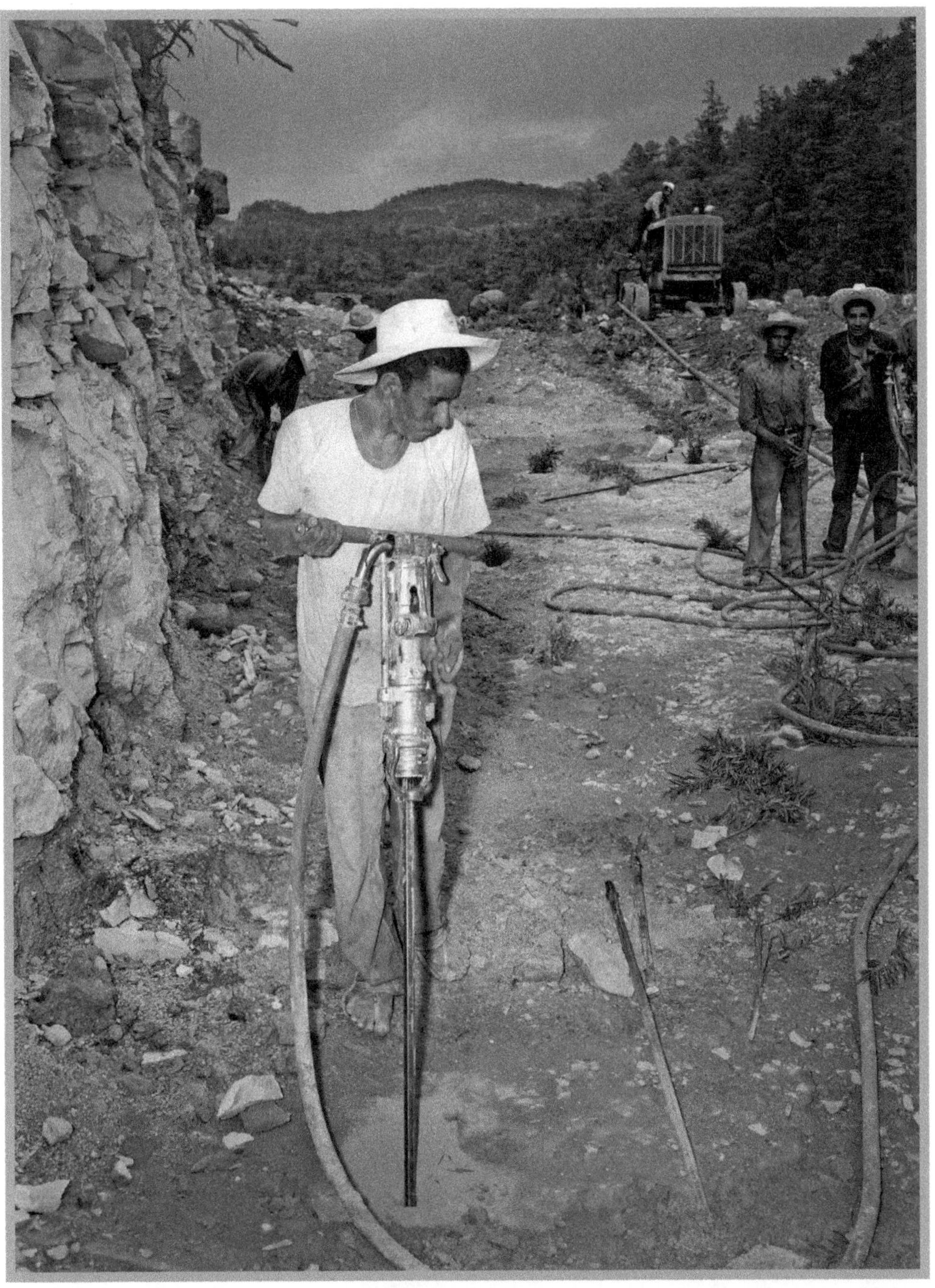

7.5. Working a compressed air drill in preparation for dynamite. Note the rubber tire sandals (*huarachis*) (photo G.B. 1955)

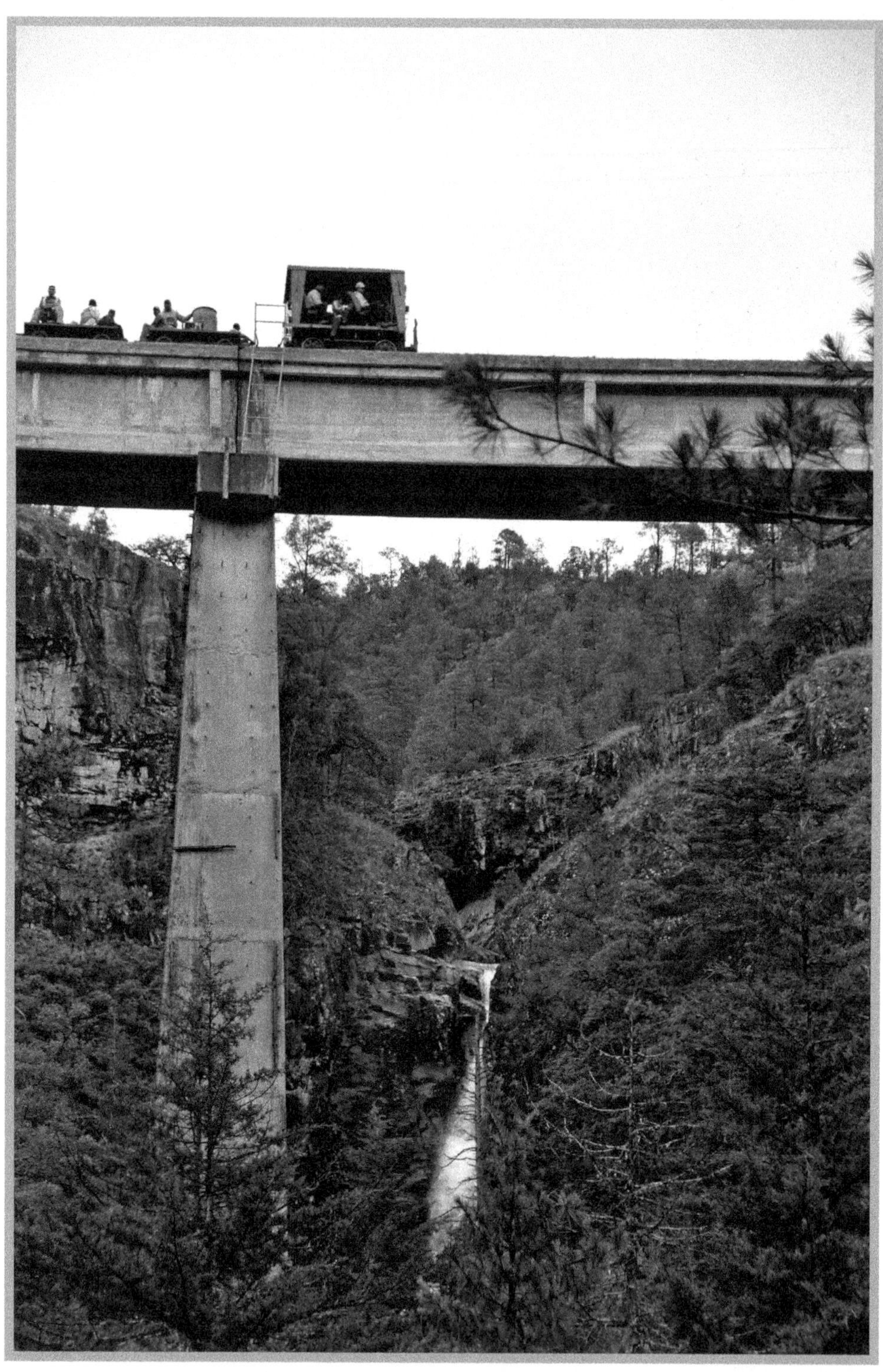

7.6. Work crew crossing bridge upstream from Cuiteco (photo D.B. 2010).

José Miyamoto

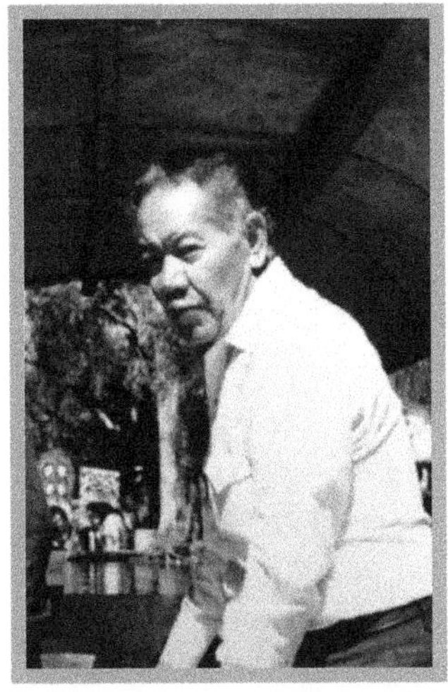

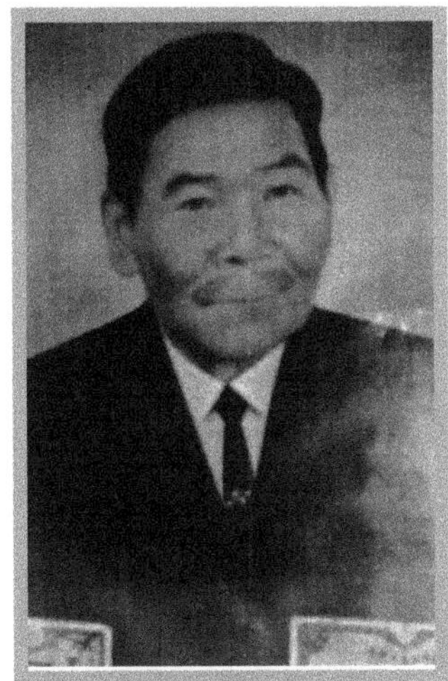

7.7 and **7.8.** José Miyamoto (c.1995) and his father, José Miyamoto Isida. Photographers unknown.

This interview was conducted in 1995 at Bachámuchi, Chihuahua, a sawmill in the mountains near Rocoroibo, where José had a store. He was born in 1923 and died July 19, 2001 on his wife's birthday. His wife is named Ana Feliz Contreras Meráz. His son Carlos continues to operate the store.[1]

| | | | | | | | | | | | | | |

Don--Tell me about the history of your dad.

José--The history of my dad? From the time he came from Japan? O.K. After the Japanese-Russian War of 1905, Japan was very poor. They had been fighting in Manchuria and they lost the war, and they had to pay restitution for the damage they had done there. So the Japanese government released from duty the soldiers and sent them to other countries to work so they could send money back to Japan. And my dad was sent to Mexico.

My dad was from Tecomamoto, close to Yokohamo, and together with 600 Japanese he came in a boat to Salina Cruz, Mexico.

1 The Miyamoto photos are courtesy of the Miyamoto family.

Don--And how many days were they in the boat?

José--It took them 45 days in the boat to get to Salina Cruz, except that they stopped in Hawaii to unload cargo. Then from there they went to Manzanillo. There they also unloaded a little cargo and from there they went to Salina Cruz. There they had to work on a sugar cane plantation, but they did not like the work because there were insects that bit and caused sores, and worms came out of those sores. So they decided to leave. They went to Mexico City. There were six Japanese who left, relatives of each other. At one point they had to cross a large river. They were on foot, fleeing. They had the idea of going to San Francisco, CA to get on a boat there because it would be impossible for them to get on a boat in Mexico. So they were on foot, walking avoiding the soldiers because the soldiers were looking for them to return them if they found them within a certain distance of the plantation. Then, when they were crossing a big river, one of them, one of the six Japanese, went into the river to check it out to see how deep it was, and when he got in the water right away a crocodile grabbed him and pulled him away. The others stayed there until the next day. And on the next day they found the body, but with one leg missing. They waited until night and at a bridge the guard was not attentive, he fell asleep, and the Japanese crossed the bridge. Then they walked along the railroad tracks.

So they arrived at Mexico City and after a time they went to Juárez, and money was sent to them there. It was sent from Japan for them to get on a boat. But they were not very wise, and they began to play roulette and other games that were there, and they lost all the money. So they had to return to Chihuahua City to work because, being Japanese without any money, they were not allowed to cross the border into El Paso. So they were there in Chihuahua City for several years working. Then from Japan they were sent a letter calling for them to return to Japan because of a war Japan was getting into [World War I], but because of the Mexican Revolution they could not leave. Then Japan sent them a card, a red card that they sent to all the men who had been soldiers in Japan, and if they did not comply within a certain amount of time, then they would be considered deserters. Then they would never be able to return to Japan. So my father could not return. Even later, when he had money, he could not go because he was considered a deserter. So, he married here in Mexico.

Before the Mexican Revolution [of 1910], they had a little store in Chihuahua City on Morelos and Independencia and it was a good place. My dad said that they even had wines and liquors in that store. It was more or less alright, but the Mexican Revolution came along and from one moment to the next they had to leave and hide, because they were foreigners and Villa was after all of the foreigners. They had no time to get all of their money: they left everything there

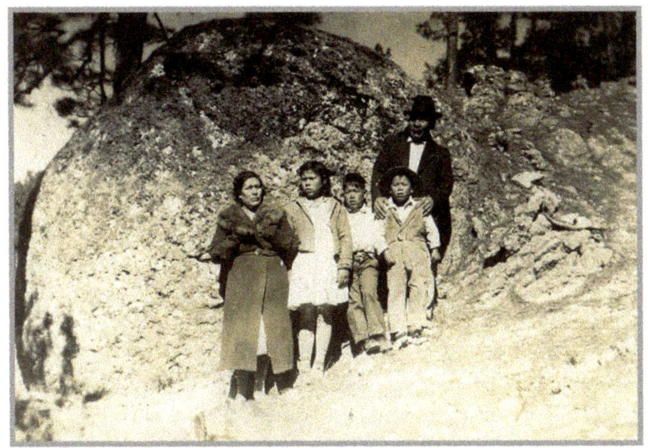

7.9. José Miyamoto Senior and his family possibly taken in the 1930s in the mountains of Chihuahua (photographer unknown).

and the revolutionaries took everything.

Then they went to hide there in the little town called Aldama, near Chihuahua City. And there the inhabitants already knew them. And every time the revolutionaries would show up, the people would hide them. The people there were good people and protected them. They were there for several years. They lived as farmers and would farm a piece of land, giving half of the crop to the owner of the land.

Then after the revolution, they went back to Chihuahua City and set up another store. Then around 1922, they got married. One of them was my father's nephew and he married first, and then my father. I and several of my brothers and sisters were born there, and then more or less in 1934 we went to the mining town of Maguarichi. I was 12 or 13 at that time. Just one of my brothers was born in Maguarichi.

We were there for several years, until the mine was not being worked any more. Then we moved to the mining town of Monterde (across the Oteros Canyon), and there we all began to work. I was about 15 years old. We worked there until the mine was closed, somewhere around 1945. My dad worked as a carpenter. In Maguarichi he had a store. It was one of the better stores there. I worked in the store when I was little. I helped my dad and uncle. They had the store together.

So from Maguarichi we went to Monterde. My dad worked as a carpenter and I worked in the mine. I worked as a common laborer. I worked inside the mine for about four years. Then there was an opportunity to work on the surface and I took that work. I worked in all the different departments. I was a common laborer, a breaker of rocks, a grease monkey, and then there was an opportunity to work with the compressors that did jobs on the surface. At this last job, I worked until the mine closed. I worked more or less six years at this mine. Then my dad went to San

Rafael. They used to call it Los Táscates (The Juniper Trees). And there he worked with the railroad engineer Togno—Francisco M. Togno he was called, right? He worked there with him. Then when Ing. Togno left San Rafael, my dad stayed there in the houses of the engineer. He gave them to my dad because they had known each other for a long time (A younger brother, Alejandro, still lives there and has a mechanic shop.) He had known my dad in Monterde when he (Togno) had been looking for a route to put the railroad. He was checking the route from Monterde to La Finca near Uruachi and from there downriver to Chínipas. But it didn't work out to put the railroad there. Togno also checked out a route that would go through Churo and down into the Septentrión Canyon, but finally, he liked the route of San Rafael, Bahuichibo, Rillito and Témoris.

We were working there in San Rafael in the tunnels near there. The work was always just for certain periods. The government just put up a little money for each year. That money would be used and the workers would be waiting for the next amount to be approved. The work would usually begin around February or March. The work would last about four or five months, and for the rest of the year the workers had nothing to do. This went on until López Mateos became president, then he got the money to finish all of the final work, until the work that was going on in the lower region met up with that of the higher region.

We worked with wheelbarrows and dump trucks. The trucks were loaded with shovels. There was no big machinery, just compressors for drilling. There were no mechanical shovels until López Mateos became president, then a lot of machinery was brought in—mechanical shovels, and bulldozers and powerful machinery. At first, we worked in the tunnels with head lamps, just like in the mines—carbide lamps. There were some light generators. They were put inside some of the tunnels, but in other small tunnels, we just used carbide lamps (*cachimbas*). That's what we worked with, and wheelbarrows which were filled by hand. At times, there were 20-30 men going with the wheelbarrows, dumping the dirt in the right place. That is how we worked at first.

I worked with the compressor, and also as a sharpener of drills. When I worked there with Ing. Togno, I worked with the compressor, as sharpener of drill bits, and fixing the wheelbarrows that would get broke. I worked in the shop (*taller*) also, when there was no need in the tunnels, fixing wheelbarrows.

Don--In what year was this?

José--The mine at Monterde closed more or less towards the end of 1945. Then we went to work in San Rafael around 1948. I am not sure.

Don--Did you live in Los Táscates?

José--No, I lived in Aremúibo, in Monterde. That's where I had my family, but I went with my dad who was in Los Táscates. I would be there a week or two, working, and then I would go to see my family in Monterde. I had married towards the end of 1945. I had a place for planting there, and I bought a piece of land and put some apple trees. I bought some animals and raised cattle, so I had something to sell, just a little, and later I had goats; I had a small herd of goats that my family cared for. I was working on the railroad bed (*bordo*). And that's the way things were.

In Cuiteco I was working there in the railroad bed and I began to work in the tunnels as a compressor operator and I was also the one who set off the dynamite when the drillers inside the tunnel had the charges connected. I was the one who had the lock for setting off the dynamite. I was the only one who had the lock and the key because it was very dangerous. When all of the machinery was out and all the drilling people were out, then I was the one who set off the explosion. And after the explosion was set off, I was the one who had to go to connect the ventilator. The ventilator pulled out the gases from the explosion. Then the machinery could be taken into the tunnel.

Don--It was dangerous, right?

José--Yes, the explosion was dangerous because it was set off with electricity, the electricity of the electric generator, and someone had to be watching carefully. I was watching the switch. I had the switch with the lock. When everyone had left the tunnel, they told me it was okay to set it off. Then I flipped the switch. I wouldn't do it until I was sure that all of the machinery was out, everything that could be destroyed.

Don--And nothing ever happened?

José--No, nothing ever happened, thanks to God. Nothing could happen with me having the lock and the key. I was the only one who could set it off.

Then an opportunity came. There was a need for a sharpener of drill bits, and as I knew how to do that well, one day I said to the engineer: "If you pay me a certain amount, I will sharpen the drill bits," because they were having problems, they could not find anyone to sharpen the drill bits really good. If they were not done just right, they would wear out quickly. It was difficult to do a good sharpening. I said to the engineer: "Pay me so much and I will sharpen the drill bits good." I told them to just send for a certain apparatus used for sharpening the bits and when it came, I fixed it up. All I had to do was push the handle where the bit was and it quickly sharpened the bit. I did so well that an inspector from the factory that made these apparatuses congratulated me because I sharpened the bits so well. He said to me: "You are one of the few who have succeeded in sharpening a bit so precisely."

I stayed there sharpening bits, and I made a little more money. A compressor operator made 20 pesos a day back then, but with this opportunity that the engineer gave me to sharpen bits, I was making 150 to 175 pesos a day, but I worked from seven in the morning until midnight. I had more than enough work, but I had to sleep also. I worked two turns at a time. 150 pesos in those days was a lot of money. The operators of the tractors didn't make more than fifty, and they worked long and hard too.

Don--I only made 17 pesos a day in Cuiteco.

José--Just think of that. I made a lot of money and I went to Chihuahua to buy a piece of land there and I made a house, and I took my family there since there were not good schools in the mountains. When I took them there, some of them wanted to study, others did not. Now I have two kids who are engineers, and a daughter is a dentist. She has a good office there in Chihuahua City, in the street Freno #710. That's where she has her office, her work. She lives more or less well.

Don--You must have been in Cuiteco in 1959, 1960, right?

José--I think so because I left Cuiteco in 1962, when the work finished. The work finished for me because there was no more blasting. There were no compressors, the machinery was being taken away, so I remained without work. They wanted to take me to another job at Infernillo, that's where they took the machinery, there in Michiocán. A dam was being made there, I don't remember the name, but a large dam was being made there by the same company. The engineer wanted to take me there.

Don--Which? The ICA company (*Ingenieros Civiles Asociados*)?

José--It was the ICA. It had a lot of machinery, and a lot of this machinery was taken to La Infernilla, and the engineer said: "Let's go there. You can keep working there." But I had my family here, all the kids and things. It was better that I stayed in Chihuahua, to work there.

Don--But you did go to Mexico City to see Ing. Togno?

José--Yes, I was working with Ing. Togno, mostly with the Ing. Monterrubio, but they were partners. At first, they were partners there in San Rafael. I was working

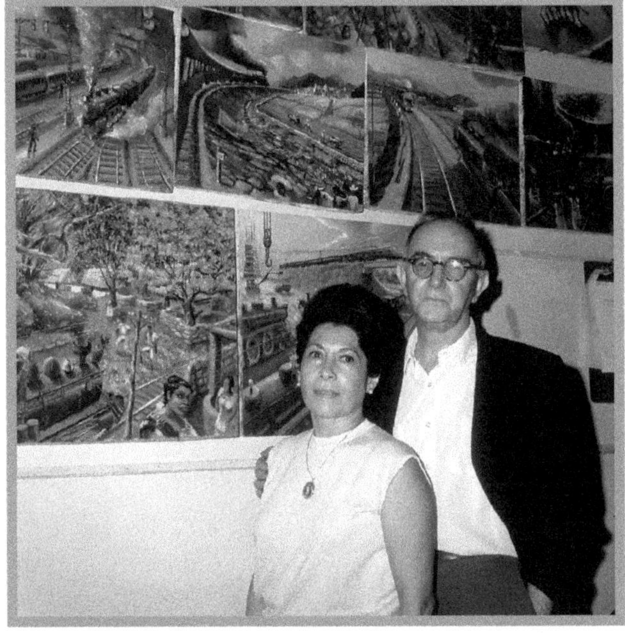

7.10. Francisco Togno and his wife Carmen Murguía Barrundia in their Mexico City home. They had five children—Carmen, Jorge, Francisco, Virginia, and Olivia. The middle three were born in Chihuahua during the construction of the railroad (photo G.B. c.1965).

there for a year, from May to May, but the machinery where I was working belonged to the Ing. Monterrubio.

Don--What did you do?

José--We made a side line for the Ferrocarril Mexicano that went from Veracruz to Mexico City. Instead of entering directly into the center of Mexico City, we made a track that went to Tlalnepantla. And we dynamited a hill that was there in a town called Ticomán. It is in one of the outlying parts of Mexico City. And we made a section also for the highway that entered at Cuauhtepec. The bus left Guadalupe for Cuauhtepec, and went through Ticomán. There in Ticomán is where we took down that hill. Up to that point I was with the Ings. Monterrubio and Togno.

Don--And did you go to the house of Ing. Togno once?

José--Yes. Once I was in the house of Ing. Togno. He received me very happily. We talked there a long time. I left in the afternoon. I had to go back to work.

Don--Was he a good person?

José--Very good person. A very fine person. I knew two or three of his kids—Panchito and a daughter, I don't remember her name. His wife was also very attentive, very good people all of them. It was the same in the house of Ing.

Monterrubio. I was there in his house several weeks because the machinery had not arrived, all the things that were needed to do the work. I was in his house until it arrived. It was on Rébsamen Street [Colonia del Valle] where Ing. Monterrubio lived. And Ing. Togno lived on Aida Street. I don't remember the number.

Don--And then when did you go to the sawmill at Lagunitas in the mountains of Chihuahua?

José--Let's see.... I took my family there. One day I was in Chihuahua City looking for work, and I saw that an engineer was having trouble standing up some large iron posts and iron vigas. He had several peons there and all the iron posts were scattered on the ground. It was there in the Deportiva. Then I said to the engineer "I will help you stand up those iron posts, but you have to pay me so much money." "Do you know about these kinds of things," he asked me. "Yes," I said, "I worked there on the railroad construction and I know more or less how to stand those up." "Alright", he said, "let's see what you need?" "Look," I said to the engineer, "bring me just a tube and a piece that is two and a half or three meters longer, because the tubes are six meters long. We needed about three meters to stand up the post. He went to the hardware store and brought these tubes and other things that we needed. I told him we needed some cables also to stand up the tubes, with four cables and some anchors for them to be firm, and they would not lean or fall. So we put the anchors on four sides so they would not fall. And we put the tube there and we began to raise the iron bars that were so heavy. We put up those iron bars and then began to put up all the building that was prefabricated in the US. There was also an iron staircase that was brought and we put together there. We were three months putting up this building and on top we put an apparatus of radar and an electrician came and connected it up. Then we put a covering of glass fiber. It was like a ball and inside this ball was the radar and I made the floor from plywood. It turned out really good.

And from there we went to make another one in Monterrey. We were there about four months also. And from there we went to Guaymas to put another.

An American was in charge of each of these buildings. We were told that these were for high atmosphere studies, but who knows. Probably it was connected with the fear that the Americans had that Russia was going to send a bomb. Only God knows, but I did work in this, I put up the buildings that the radar sat on. There was a large amount of apparatuses there.

Don--When did you go to Lagunitas?

José--When we finished these buildings, the last one being in Guaymas, I went

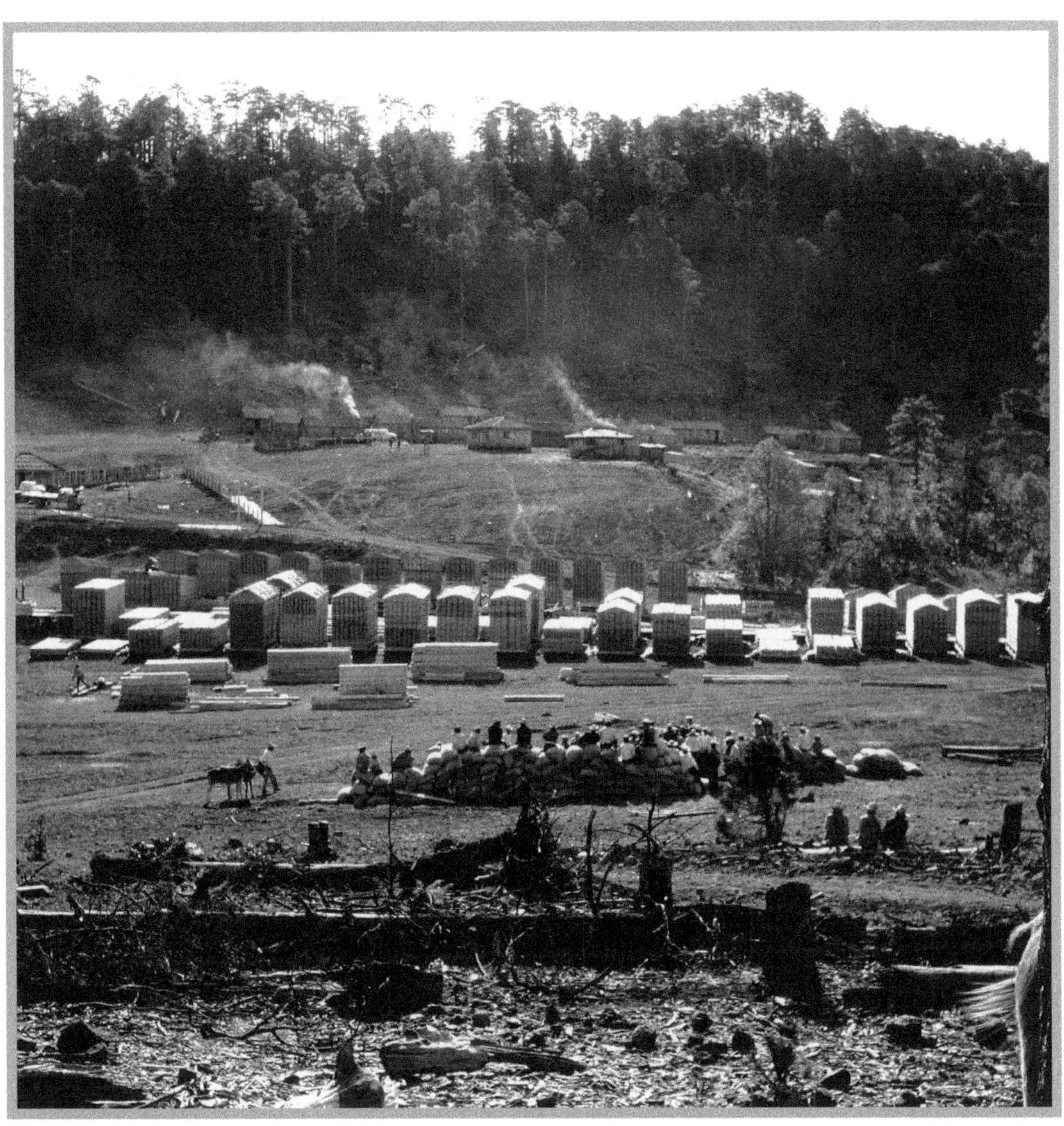

7.11. Lagunitas sawmill where José had a store. A few days after this photo was taken, the lumber seen in the photo was burned by rebels. Bags of corn can be seen in the foreground, payment to local people for lumber (photo D.B. 1974).

to Chihuahua City, and I heard that they were going to put a sawmill here in Lagunitas, here on top of the mountain and I went to see because I had a desire to put a store there. So I went there and began to make a house. I had credit with the Cuestas in Creel, Gregorio Cuesta, and they would stock me. I had known Gregorio a number of years.

Don--Did he send food to San Rafael for the railroad workers?

José--No. Gregorio had stocked several stores there in San Rafael and that is where he knew me and my father. So I asked him for credit. I told him I wanted to put a store in Lagunitas. And he said to me: "Sure, go for it, no problem. Just make a house and get settled there and I will see how to stock you." Back then express was sent on the train, so he sent mercantile from Creel to San Rafael. I received it there and brought it out by truck.

Don--What year was that?

José--I am very bad for dates. I don't remember exactly.

Don--So you put in a store.

José--Yes. I was working there until those rebels came who burned the sawmill [1974], then I changed from there to here in Bachámuchi.

Don--And what was it like when those rebels arrived? Did they arrive at your store?

José--Yes they arrived there, but just on that day they burned the sawmill. Before that I did not know them, but this day they arrived there where I was because I was commissioner of police there in Lagunitas and they wanted to get Enrique Sonoscar [Oscar?--the head of the sawmill] because he had promised higher wages and I do not know what else. And since they did not find him, they grabbed several persons, but they took me prisoner and had me in front of the sub-machine guns. They were saying things there about what they were after and wanted, and then the "jefe" of them said, "We are going to burn this sawmill so that people will know we mean business." They said they were against the rich people, and I don't remember all the other things they told me. And I said to them that they should not burn there where the poor people had their houses. "They only have the little that they have there in their houses. So please do not burn there where they live." So they burned the sawmill, but not the houses of the poor people.

Don--What was your father called?

José--My father was called José Miyamoto just like me. Except that his second family name was Isida. In Japanese that means corral or where they plant, or where it is planted but there are lots of rocks.

Don--And Miyamoto, what does that mean?

José--I don't know.

Don--Does it mean flower?

José--No. Back to Isida, it means "field with rocks," or "rock in the field", something like that. There where it is planted.

Don--And your father is buried in Areponápuchi?

José--Yes, he is buried there. He said that that was where he wanted to be buried. He was 85 when he died. I am 72. Day after tomorrow I will be 72. Day of the Dead. On the Day of the Dead I will be 72.

Now we have talked some.

I remember a Japanese story: It says that, you will see...my father said there was a married woman that said: "How come I do not have eyes on the ends of my fingers in order to look for things up high, like above the wardrobe (*ropero*) and the dish cabinet (*trastero*)? But then another woman said to her, because another lady was there in the kitchen grinding chile, "Why would I want eyes on the fingers if I am sticking my hands in the chile."[2]

2 When José told me this story once before, the cook was working with a hot fry pan instead of chile.

Effects of the Railroad on Local Cultures

by Don Burgess

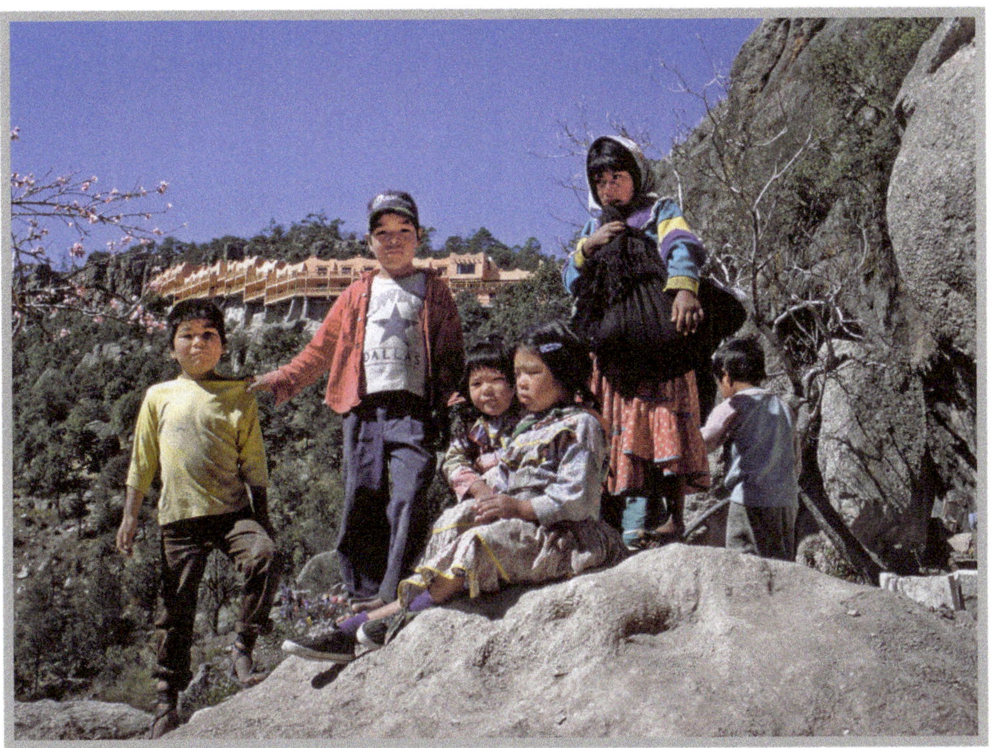

8.1. Tarahumara children who live in cave homes below a modern hotel near El Divisadero (photo D.B. 2005).

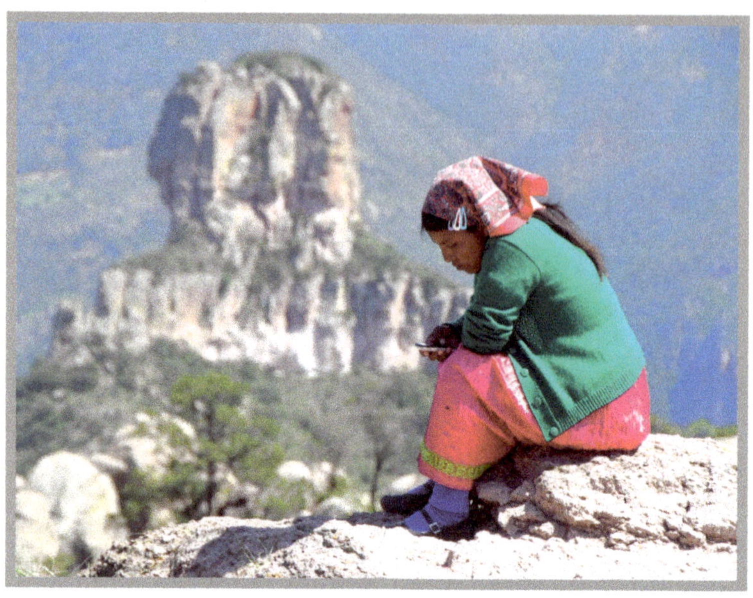

8.2. Tarahumara woman with cell phone in Urique Canyon where the cable car "Teleférico" ends (photo D.B. 2012).

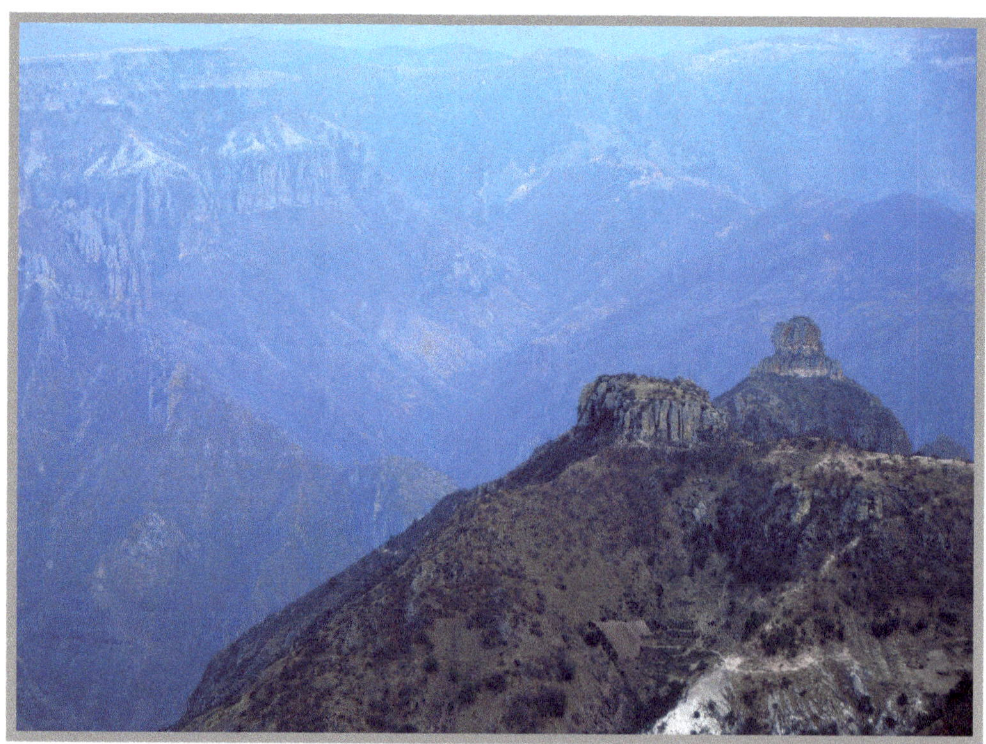

8.3. The flat-topped rock (mesa) in the mid-center is where the woman is sitting to make a phone call in the top photo. The lower photo was taken a few years before cable cars were installed that go from one of the Copper Canyon overlooks to this flat area, half way across the canyon (photo D.B. c.2005).

> The Chihuahua al Pacífico Railroad will be, without doubt, an element of extraordinary importance for the economic and social development of the Tarahumara. They will be able to use resources that until now have existed only in their potential form, and if the beneficial action of the railroad is successfully channeled for the benefit of the communities and of the Indian communal counties (*ejidos*), certainly it will serve as a powerful lever for the economic and social benefit of these communities. (Lic. Alfonso Caso, Director of the National Indian Institute, in *Comunicaciones y Transportes*, 1962, p.19.)

The two cultures most affected by the construction of the railroad across the Sierra Madre Occidental are the Tarahumara and the local "Mestizo."[1] Both cultures have faced rapid changes brought about by the loss of an isolated environment.

The Tarahumara area covers about 30,000 sq. miles (Before the Spaniards arrived it covered approximately 45,000 sq. miles.) and the railroad cuts right through their territory. It has always been difficult to find precise information on the Tarahumara population, but the figures vary from between 55,000 to 85,000 people (Juan Luís Sariego, p.13; INEGI 2010 Census).

A little history is in order. The Tarahumara, or Ralámuli[2] as they call themselves, are part of the Uto-Aztecan linguistic group which includes the Utes in the northern United States, the Hopi, Pápago-Pima, Yaqui, Mayo, Guarijío , Tepehuán, Cora, Huichol, Aztec, etc. throughout the length of Mexico and into Costa Rica. The closest linguistic relative to Tarahumara is Guarijío, followed by Yaqui and Mayo.

Linguistic comparisons suggest that the Tarahumaras have been in their present area for several thousand years. There are at least two major dialects of the language (called the Alta and the Baja), whose speakers have difficulty understanding each other and for which the government is making separate school books. Within each of these areas are numerous dialect differences. The canyons (*barrancas*) and distances have kept the people from being a more unified group both linguistically and socially. They live a very scattered, semi-nomadic "rancheria" existence and have been able to stay out of the mainstream of Mexican life until this past century.

1 The Mayo Indians (Yoreme) who live more towards the coast have also been greatly affected by the arrival of the railroad. In fact, they have seen the operation of the Chihuahua al Pacífico in their area since the line from Topolobampo to El Fuerte was completed in 1904 (Kerr, p.74).

2 The Tarahumara r is a "retroflexed, forward flap" r. The tip of the tongue curves back and flaps forward, the opposite of the Spanish r. The l is also retroflexed. Some say that the word Ralámuli means "foot runners," but that is most likely not correct. There are other linguistic possibilities.

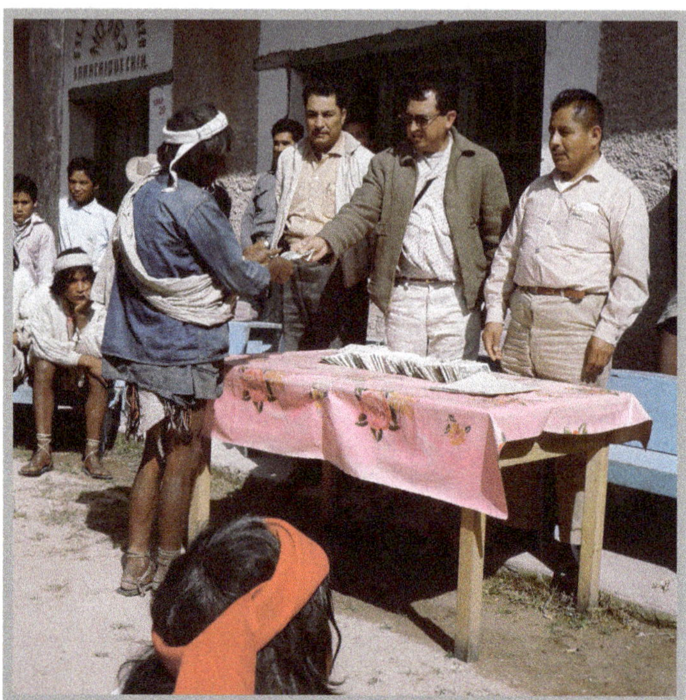

8.4. Government officials giving out money to Tarahumaras in Samachique in the 1960s as part of profit sharing for the exploitation of the forest (photo D.B.).

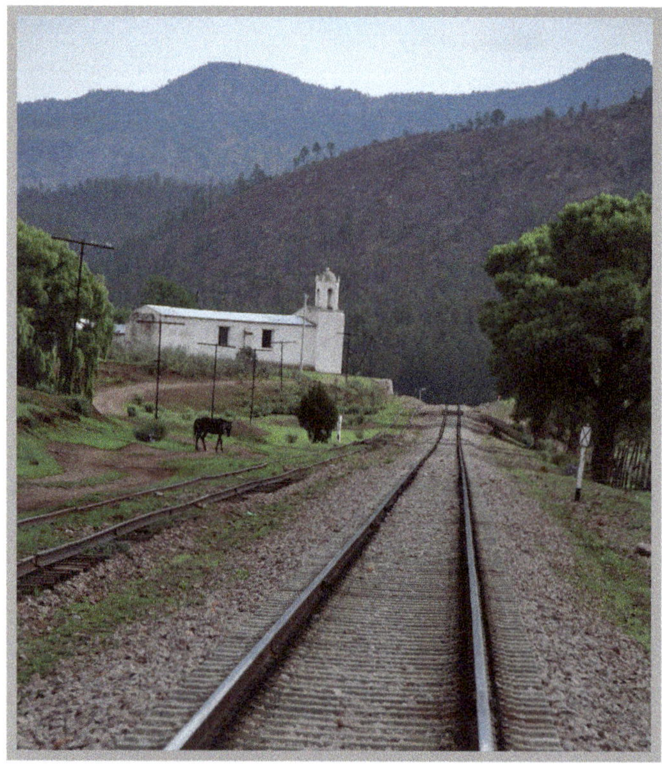

8.5. The railroad passes by the Catholic church at Pichachí. The first mention of this mission in the Jesuit writings was 1678 (photo D.B. 2012).

Almada writes that the first Spanish forays (*entradas*) into the Sierra Tarahumara came from the Pacific-Sinaloa side of the mountains in 1598 with Spaniards searching for minerals in the Chínipas region. The founder of the Tarahumara mission movement was the Jesuit Pedro Méndez who went as far as Cuiteco[3], along with Captain Diego Martínez de Hurdaide, 23 soldiers, and some Indian guides. They were attacked by Tarahumaras a few miles below Cuiteco. Several Tarahumaras were killed and a few of the soldiers were wounded. They must have come up the Septentrión Canyon, at least parts of it, near where the present day railroad runs.[4]

On the other side of the Sierra, the Jesuit Juan Fonte began mission work in the area of Parral, coming up from the Tepehuán missions in 1604. He was killed in the Tepehuán revolt of 1616, led by Quautlatas. The missions spread northward from Parral beyond La Junta into the Papigochi area. The Tarahumaras living in this area assimilated into the Spanish culture or retreated into the mountains where the other Tarahumaras lived. Tarahumara place names in this area, such as Papigochi, are still used.

The last half of the 17th century was marked by several revolts by the Tarahumaras against the Spaniards such as the 1650 [1652] revolt led by Tepórame. In the late 17th and early 18th century, the Spaniards established themselves in mines such as Cusihuiriachi (1687), Batopilas (1708)[5], and Urique (1690), and in areas that were good for farming. The Tarahumara revolts during those years sometimes involved Tarahumaras joining with Apaches.[6]

The Jesuit mission system flourished during this time, but in 1767, the King of Spain expelled all of the Jesuits from the Americas, and 19 priests had to leave the Tarahumara area. Most of their missions were turned over to the Franciscans from Zacatecas, who were there for a number of years before moving on to California.[7]

The Jesuits returned to the Tarahumara area in 1900, and, soon after this, protestant missions began to work in the area as well (Don Burgess, "Missionary

3 The Tarahumara pronunciation of Cuiteco is Gutego, which refers to throat, or a narrow place which then opens into a large area. Tarahumaras sometimes name their ranches according to geographical characteristics which look like body parts (See Don Burgess, "Western Tarahumara Place Names," pp.65-88).

4 See Francisco Almada, *Resumen de Historia del Estado de Chihuahua* and *Apuntes Históricos de la Región de Chínipas*; and Luís González Rodríguez, *Crónicas de la Sierra Tarahumara.*

5 See Grant Shepherd's book *Silver Magnet.* Dates for the discovery of these mines are from Almada's *Diccionario de Historia, Geografía y Biografía Chihuahuenses.*

6 See William Merrill, *Raramuri Souls* (pp.34-35), and Don Burgess, "Missionary Efforts Among the Tarahumara Indians" (thesis) (pp.38-56).

7 For more on the early history, see Peter Masten Dunne, *Early Jesuit Missions in Tarahumara*, Campbell W. Pennington, *The Tarahumara of Mexico (Their Environment and Material Culture)*, and Luís González Rodríguez, *El Noroeste Novohispano en la Época Colonial*, and *Crónicas de la Sierra Tarahumara.*

Efforts Among the Tarahumara Indians", pp.57-80.) and the government began to take a more active role in Indian affairs. The radio station XETAR, The Voice of the Sierra Tarahumara, that transmits in Alta and Baja Tarahumara, Ódami (Tepehuán), and Warijío (Warijó), celebrated its 30th anniversary in November of 2012.[8]

It can be seen that the question of how the construction of the railroad has affected the Tarahumara people is part of a much larger picture. It has to be seen in the light of the Spanish "entradas" into the Sierra Madre Occidental, the different mission efforts, the mines, the arrival of illnesses such as TB, the government programs, the lumbering, the schools, tourism, and the planting of marijuana and opium poppies. In addition to the intrusion of lumbering and mining roads, competing with the railroad to open up the Sierra, new developments include a paved road nearing completion that will connect to the Pacific, an almost completed jet port near Creel, and a three kilometer long cable car ride which goes down into the Urique Canyon.

Tarahumaras themselves mention the inadvertent effects of transportation developments such as animals killed by trains and grass fires caused by sparks and cigarettes. Our Tarahumara friend Antresi, who often brought us wood, has several fingers missing from when he was drunk and was lying next to the rails when the train went by. Another Tarahumara friend tells me that his younger brother and some other Tarahumara boys from San Luis Majimachi jumped on one of the small railroad cars called "armones" when the workers were not looking, rode it for a couple of miles to Nacárare, and then jumped off while the car continued down the track.

The loss of an isolated environment necessarily meant that Tarahumaras have had to change. The Tarahumara culture had been affected by the Spanish culture long before the railroad entered their territory. The following by Luís Urías is a partial list of everyday objects, plants and animals found in Tarahumara life that came from the Spanish culture: iron tools, metal objects, horses, donkeys, mules, goats, sheep, chickens, pigs, wheat, onions, oranges, limes, sugar cane, spices, medicinal plants, cows, milk, cheese, butter, and musical instruments, as well as melodies, rythms, harmonies, symbols, and customs (as found in Burgess and Mares, *Con el maiz se pueden hacer muchas cosas*, p.4). Even the Tarahumara greeting "Cuira" came from the Spanish "Dios os cuide", "may God care for you" (or perhaps "Dios os cuida," "God cares for you"). But I have seen accelerated changes taking place in the towns along the railroad. In these towns, Tarahumaras have benefited from access to schooling, more medical care, and paying jobs. But at the same time, they are losing their culture because they are immersed in a new culture and language, eating different foods, and watching television. Some of these Tarahumaras are

8 Personal communication from Luís Urías, the first manager of the radio station.

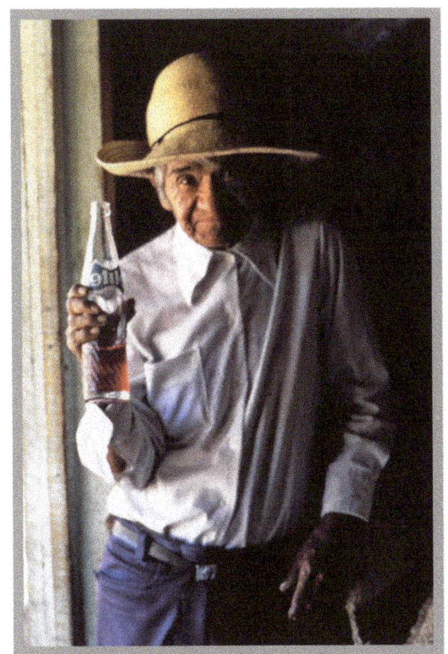

8.6. Tarahumara man drinking a bottled soda at Lagunitas (photo D.B. c.1970).

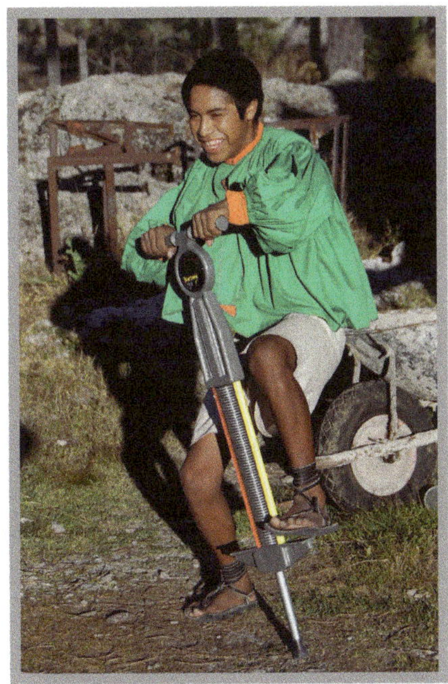

8.7. Tarahumara boy, Bernardo Jaris Núñez from the La Bufa-Batopilas area, jumps on a pogo stick (photo D.B. 2012).

even beginning to lose their language.

The effect of the railroad, however, should be considered not just from the standpoint of how outsiders are affecting the indigenous people, but also from the standpoint of how the choices of the native people are affecting the situation. Some are good choices for the benefit of the people, but some are not. Greed and drunkenness among the Tarahumaras have had devastating effects.[9]

It must be said that Tarahumaras have certainly made use of the train, although it must have taken some of them a time of adjustment. Several Tarahumaras have told me that the first time they ever saw the train was from high up on a ridge looking down to where a train was moving along the bottom of a canyon, and they thought it was a snake. Some were afraid to get on the train and might walk long distances in order to avoid the train.

The main benefit to Tarahumaras has been for transportation and for the importation of merchandise. Many Tarahumaras for years have worked seasonally cutting sugar cane, picking tomatoes, etc. in the coastal plains, and the train quickly became their means of transportation to get there. Furthermore, the train has

9 See Merrill, William L. 2001. "La identidad ralámuli: una perspectiva histórica," in *La identidad y los pueblos étnicos en la Sierra Tarahumara*, pp.71-103.

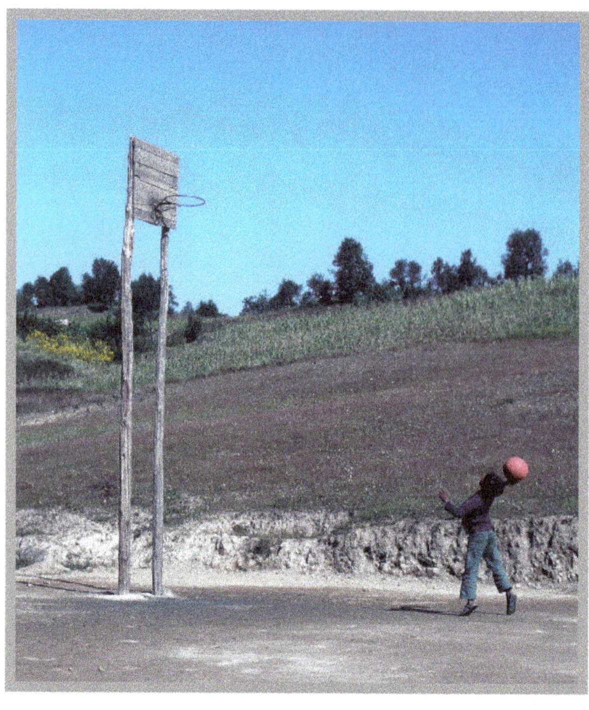

8.8. Tarahumara boy playing basketball at a government school ground at San José del Pinal. The hoop was set at 15 feet. When I asked why, I was told they did not want to cut the poles off (photo D.B. 2008).

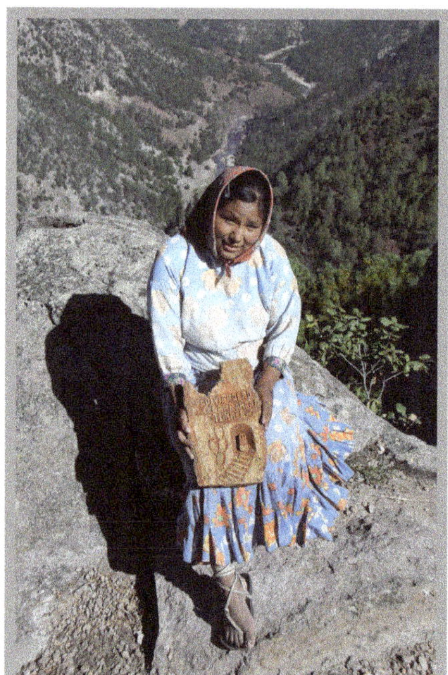

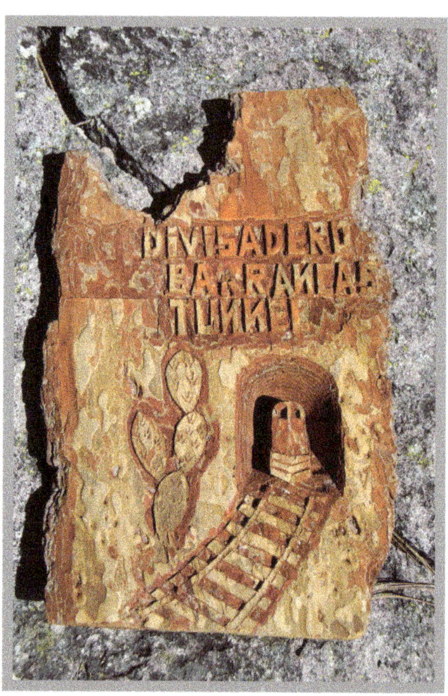

8.9 and **8.10.** Pine-bark carving of train in tunnel made by Tarahumara, Juan Cleto González, and sold by his daughter, Hortencia Cleto Cruz, at the Rio Oteros Overlook. It says Divisadero Barrancas Tunnel. The carving reflects the development of products designed for the tourist market (photo D.B. 2012).

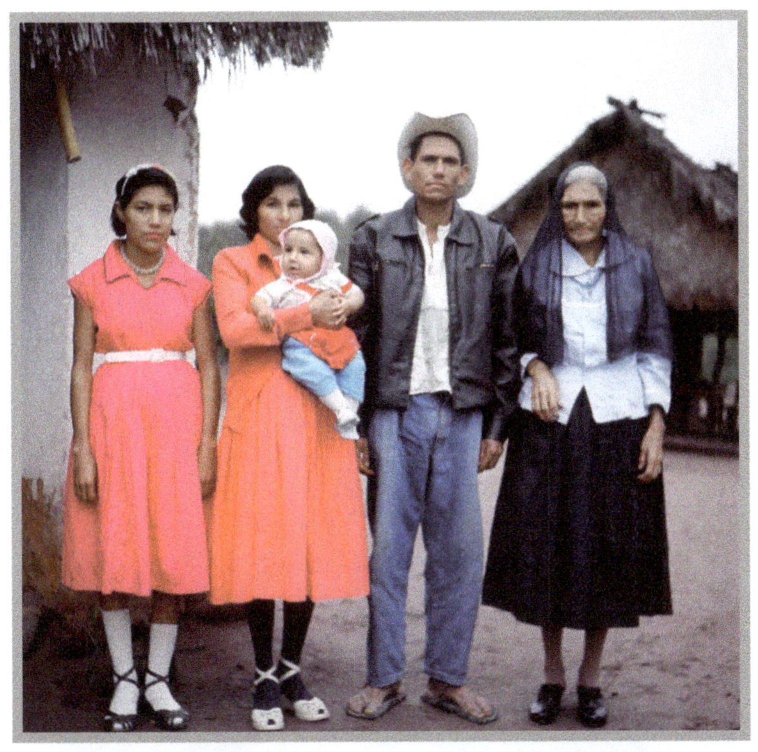

8.11. Doña Lola on the right, her son José Portillo, his wife Challa, and their kids. The Portillo family lived for several generations at Socolén and other places along the San Ignacio River, one of the alternate routes of the railroad (photo D.B. c.1970).

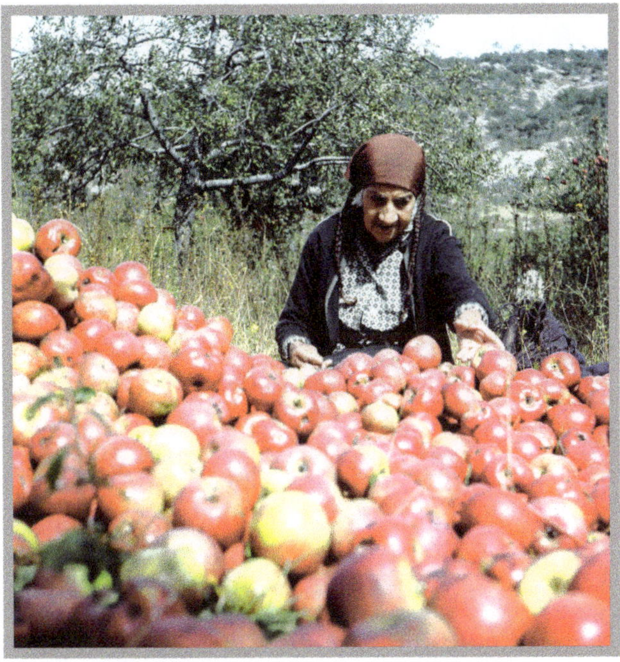

8.12. Doña Lupe García sorting apples near Areponápuchi. She was the grandmother of Armando Díaz, owner of Cabañas Díaz, and a relative of Doña Ponciana, a pistol-packing lady whose remodeled house is now part of the Balderrama's hotel at Areponápuchi. Alejandro Miyamoto remembers Doña Ponciana riding up to his house (Ing. Francisco Togno's old house) with pistols and a rifle, and her pants tucked into her boots. Her shooting skills are legendary (photo D.B. c.1975).

brought goods to local areas. I know Tarahumaras who used to yearly make the two month trip with burros to the coast to get salt. Now they only have to go to the closest store.

The long history of Spanish colonization in Mexico led to the establishment of Mestizo communities in the Sierra Madre. This local Mestizo culture (whose population today is more than twice that of the Tarahumara) has also been affected by the railroad.

Like the Tarahumara, their culture is threatened by the loss of isolation. The changes for them have been similar to those found in the passing of the Old West in the United States. An article in the publication *Mining and Scientific Press* (Nov. 4, 1911) stated:

> The completion of the Southern Pacific railroad along the Mexican west coast, with its branch to Alamos, and the building of the coast end of the Kansas City, Mexico & Orient railroad to El Fuerte in Sinaloa, marked the passing of the old order in mule-back transportation and the closing of the coast trails of Sonora and Sinaloa. The completion of the latter road will completely eliminate the mule as a factor in

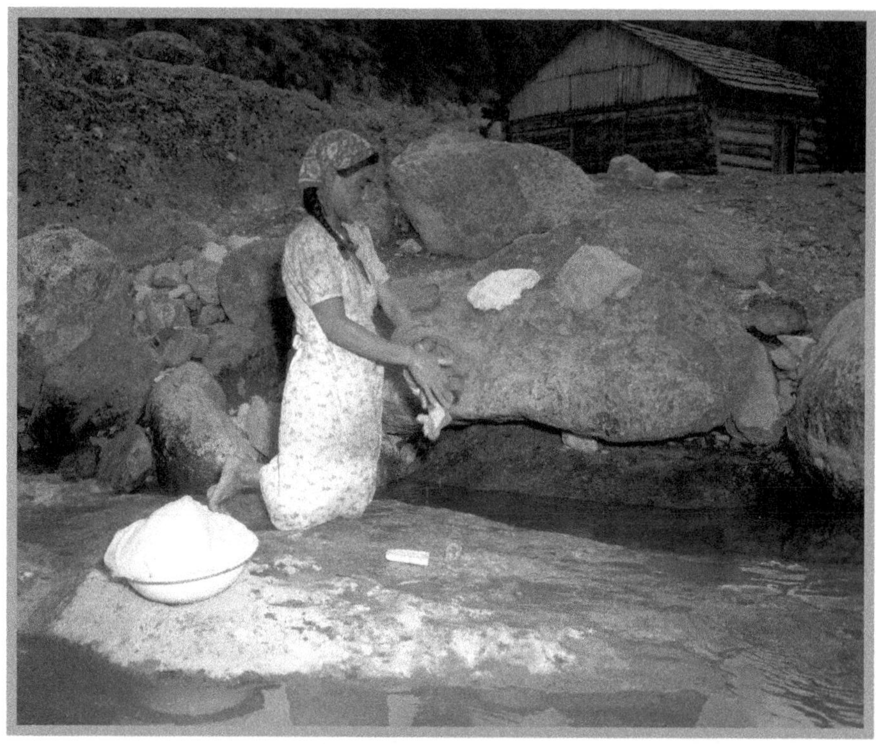

8.13. Mexican woman washing clothes in a mountain stream (photo G.B. 1955).

> freighting, and put an end to a picturesque phase of Mexican life along many historic trails of the interior.... The campfires on the hills at night; the laughing, singing *arrieros* (mule drivers), the mules and burros, feeding, fighting and neighing, are part of the memories of a great highway whose glory has gone forever.

Those were the days of the *Camino Real* (the Royal Road) as the major trails were called. One Mestizo family that I know lived on a ranch in the Tubares area. I remember Doña Challa saying, as her husband José Portillo came up to the house with a number of mules and donkeys loaded with provisions: "*El Camino Real es muy duro*" (The Royal Road is very difficult.). It had been a two day trip (one way) from the town of Choix to their ranch. Such trips were the only way that the family could access provisions. Most Mestizo families lived on such isolated ranchos, scattered across the Sierra. (I myself used a mule for a number of years to pack into the isolated valley where we lived in the Sierra.)

Those were also the days before radios and televisions were brought into the sierra. It was a time of story-telling, when there were "professional" story-tellers. People would bring food and other items to these men and listen to their tales. In my trips through the mountains and canyons, I collected a number of these "tall tales" told by the Mestizos which I put into a book called *Chistes De La Sierra Tarahumara (Contados por la Gente Mestiza)*.

I tell in that book what a Mestizo rancher who lived in the bottom of the Chínipas Canyon told me about the first radio he ever saw. Someone had brought one to their ranch and the people figured out how to turn it on, but could not figure out how to turn it off. After a while they got tired of hearing all the noise and piled blankets on top of the radio. Finally the battery went dead and the noise stopped.

Tall tales and jokes say a lot about how people perceive change. I will close with a Mestizo story about the first time two men decided to ride the train. It goes like this:

> This is a tale about two men.
>
> They were brothers and good friends. They had never been on a train and were about to get on the train, and one said to the other:
>
> "You go first 'compadre'.
>
> "No you," said the other.
>
> "No, you get on."
>
> Finally one of them got on and went and left their suitcase in the aisle of the passenger car. Then he got back off where the 'compadre' was,

and, as they continued talking, the train whistle blew announcing that the train was ready to leave, and:

"Get on 'compadre'."

"No, you."

"No, you."

Well, the train suddenly took off, and neither one of the two got on. The train left and the suitcase left in the passenger car.

Well, there was a gringo in the passenger car, right next to the suitcase, and the conductor said to the gringo:

"Get that suitcase out of the aisle."

The gringo didn't do anything. Since the suitcase wasn't his, he didn't move it. And the conductor passed by again and said to him:

"Get that suitcase out of the aisle because it is blocking things."

The gringo still didn't do anything.

The conductor passed by again, and he said again to the gringo:

"Quita el veliz...." Get that suitcase out of here. If you don't get it out of the aisle, I am going to throw it outside."

"It doesn't matter to me. Toss it," said the gringo.

So then the conductor grabbed it and threw it out the window.

Then the gringo said to him:

"Oh, it doesn't matter to me because the suitcase is not mine."

"Ahh, it isn't yours?"

"No."

And the conductor went running to stop the train. He got down, found the suitcase and put it back up on the train.

And that's how this *chiste* happened. That's the way it ends.

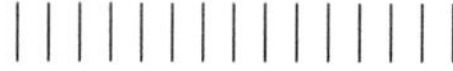

A story (perhaps based on an actual occurrence) became a *chiste*, a tall tale using local characters. This *chiste* certainly tells us a lot about the impact of the railroad on local life for the Chihuahua al Pacífico Railroad brought a new Sierra Challenge.

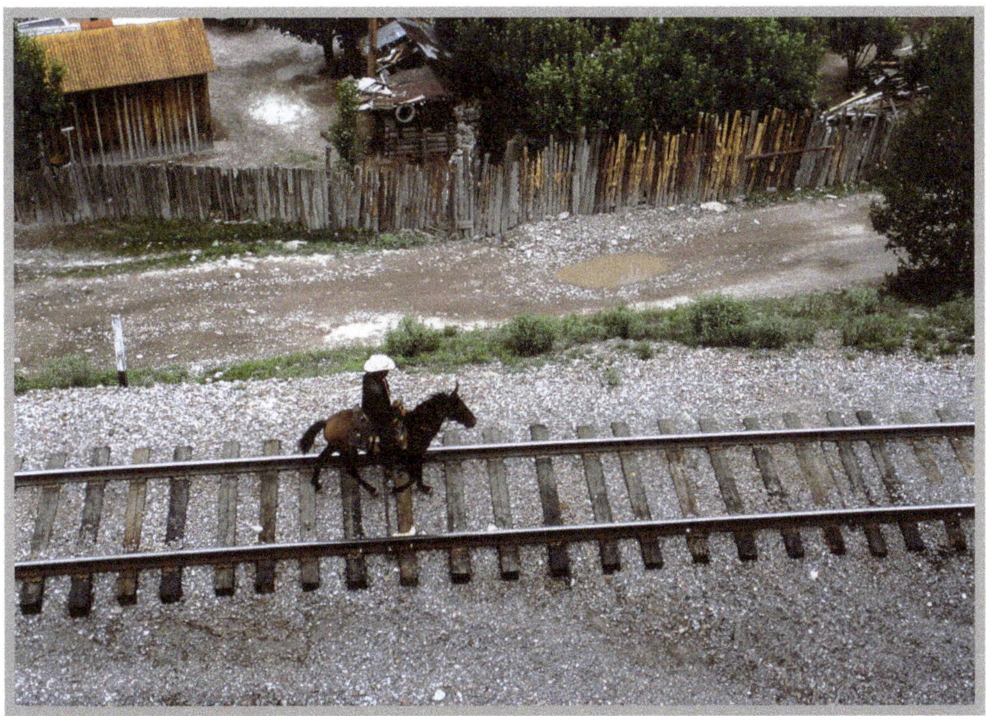

8.14. Man riding horse along railroad tracks through the town of Creel. He likely doesn't ride on the parallel road because it does not go far enough (photo D.B. 2002).

References

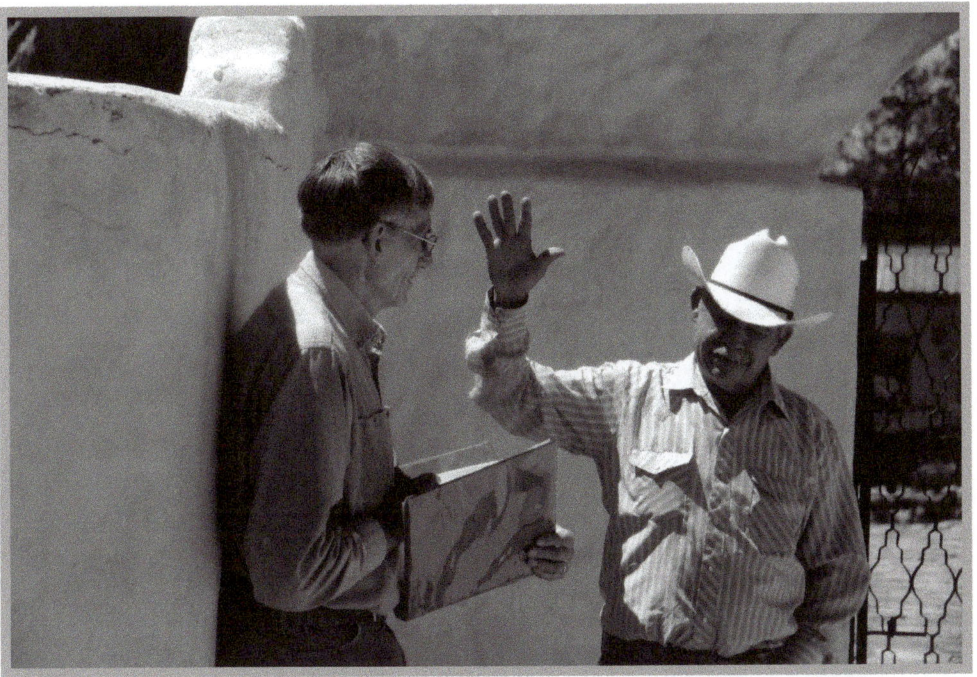

9.1. Don talking local history with Armando Díaz at Areponápuchi (photo by Marie Burgess, 2012).

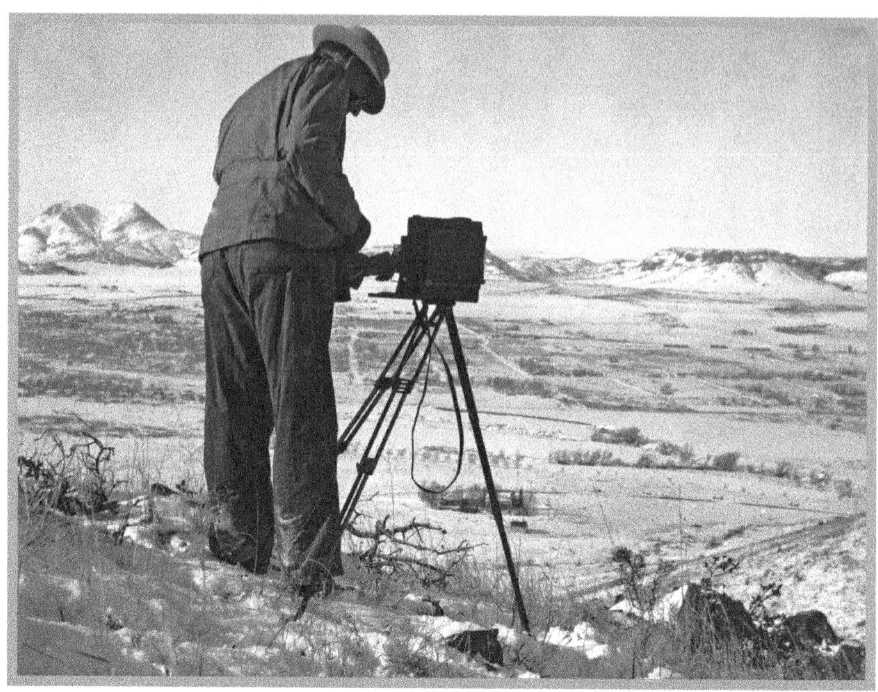

9.2. Glenn taking photos with a Speed Graphic camera near Alpine, Texas (photographer unknown c.1960).

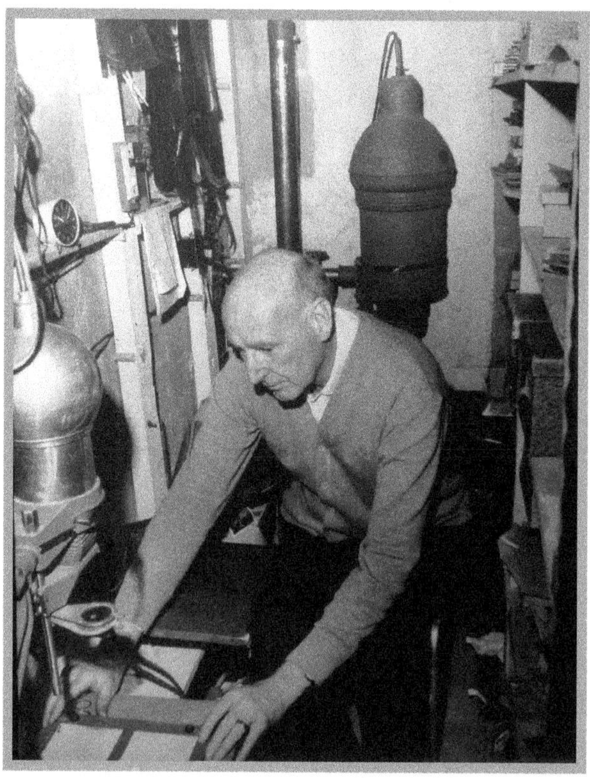

9.3. Glenn printing photos in his lab in Alpine, Texas (photographer unknown c.1960).

Railroad articles by Glenn Burgess

"Chihuahua-Pacific Railway To Open Mexican Resources." *The El Paso Times*, May 19, 1957.

"Work on Two Mountain Railways In Mexico Fast nearing Completion." *Fort Worth Star-Telegram*, August 20, 1957.

"Railroad To Unite Sinaloa, EP, Presidio." *The El Paso Times*, September 24, 1957.

"Railroads Build Mexico's Future." *The El Paso Times*, Oct. 9, 1957 (article about a proposed railroad line from Falomir to a mining area 80 miles to the southeast of Ojinaga).

"Chihuahua Railroad Forges Pacific Link." *The El Paso Times*, May 2, 1959.

"Dream Mexico Railroad Now Nears Completion." *Fort Worth Star-Telegram*, May 10 1959.

"Tortuous Terrain Shattered Early Completion Hopes of Railroad." *The El Paso Times*, May 15, 1959.

"New Railroad in Mexico to Penetrate Sierra, Do 'Impossible'." *Fort Worth Star-Telegram*, May 22, 1959.

"Tough Railway Project On Last Lap In Mexico." *Fort Worth Star-Telegram*, July 5, 1959.

"Final 30 Miles of Chihuahua Al Pacífico Railroad The Toughest." *The El Paso Times*, July 6, 1959.

“Mexican Rail Project to Tie Texas with Pacific." *Fort Worth Star-Telegram*, (date unsure).

"Long Sought Rail Link to Open." *The El Paso Times*, October 16, 1960.

"Sierra Challenge." *The CF&I Blast*, May 22, 1961.

"Chihuahua Al Pacífico Railroad Now is Startling Reality." *The El Paso Times*, January 27, 1963.

"Spectacular Scenery Yields Impact To Chihuahua Al Pacífico Train Ride." *The El Paso Times*, February 3, 1963.

References

Alegre, Francisco Javier. *Historia de la Compañía de Jesús en Nueva-España, que estaba escribiendo el P. Francisco Javier Alegre al Tiempo de su Expulsión*, Tomo III. Mexico: 1842.

Almada, Francisco R. *Apuntes Históricos de la Región de Chínipas*. Tallares Linotipográficos del Estado de Chihuahua, 1937.

---. *Resumen de Historia del Estado de Chihuahua*. México, D.F.: Libros Mexicanos, 1955.

---. *Diccionario de Historia, Geografía y Biografía Chihuahuenses*. Juárez: Impresora de Juárez, S.A., 1968 (second edition).

---. *El Ferrocarril de Chihuahua al Pacífico*. México D.F.: Editorial Libros de México, S.A., 1970.

Brambila, David. *Marko (El Evangelio Según San Marcos): Traducción bilingüe*. 1993.

Brown, Ben. "Chihuahua bound: or how I spent my Christmas in 1881." *The Chronicles of the Trail*. Camino Real Trail Association, Vol.1, No. 4. October. December 2005.

Brown, R.B. (ed.) *Introducción e Impacto del Ferrocarril en el Norte de México*. Juárez: Universidad Autónoma de Ciudad Juárez, 2009.

Burgess, Don. "Mexican Railway to Pacific Coast 95% Complete, Alpine Youth Finds." *Alpine Avalanche*, Aug. 20, 1959.

---. "History of Missionary Efforts Among the Tarahumara." Master's Thesis. El Paso: Texas Western College, 1963.

---. *¿Podrías Vivir Como Un Tarahumara?* (Fotografías: Bob Schalkwijk y Don Burgess). México: Bob Schalkwijk, 1975.

---. "Western Tarahumara," in *Studies in Uto-Aztecan Grammar*, Vol. 4 (Ronald W. Langacker editor), 1984.

---. *Chistes De La Sierra Tarahumara (Contados por la Gente Mestiza)*. Chihuahua: Don Burgess McGuire, 1987.

---. "Western Tarahumara Place Names." *Tlalocan*. México, D.F.: Universidad Nacional Autónoma de México, 1990.

---. Interviews with Leopoldo Méndez (San Rafael, circa 1980), José Miyamoto (Bachámuchi, 1995), Candelario López (San Rafael, 2003), Ing. Eloy Yáñez (Chihuahua City, 2012), Ing. Miguel Leal (Chihuahua City, 2012), José María Ávila and Rogelio Ávila (Chihuahua City, 2012), Armando Díaz (Areponápuchi, 1012), Lisa Wolf (Cuauhtémoc, 2012), Enrique Wolf (Cuauhtémoc, 2012), Jacobo Harms, Susan Thiessen and Abraam Thiessen (Cuauhtémoc, 2012), Salvador Bustillos (Creel, 2012), Antonio González (Pichachí, 2012), Carlos Jaime (San Juanito, 2012), Anacleto Ramírez (phone, 2012), Oscar Luévano (correspondence, 2012), Rosalva Delgado (correspondence, 2012), Lic. Roberto Balderrama (correspondence, 2012),

Robert Schmidt (correspondence, 2012), William Merrill (correspondence, 2012), Luís Urías (correspondence, 2012), Olivia Togno (correspondence, 2013).

--- and Albino Mares. *Sunute we'ká e'karúgame newalime ju (Con el maíz se pueden hacer muchas cosas).* Cuauhtémoc: Don Burgess McGuire, 2010.

Burgess, Don (translator). *Onorúgame Nila Ra'íchali (El Nuevo Testamento en el idioma ralámuli de la Tarahumara Baja del municipio de Guazapares, Chihuahua, México).* México, D.F.: La Liga Bíblica, A.C., 2008.

---. *Onorúgame Nila Ra'íchali Mapu Ruwime Ju Alué 'We 'Ya Osirúgame Napu Antiguo Testamento Anilime* (Resumen del Antiguo Testamento en el idioma ralámuli de la Tarahumara Baja del municipio de Guazapares, Chihuahua, México). México, D.F.: La Liga Bíblica, A.C., 2013

Burgess, Glenn. "History of the Kansas City, Mexico and Orient Railroad" (Masters Thesis). Silver City, NM: New Mexico Western College, July 1962.

Casey, Clifford B. *Alpine, Texas then and now.* Seagraves: Texas Pioneer Book Publishers, 1981.

Coffey, Fred. "Confederate Attempts to Control the Far West" (Master's thesis). Austin: The University of Texas, Austin, 1930 (as quoted in Glenn Burgess's thesis).

Collard, Howard and Elizabeth Scott Collard. *Vocabulario Mayo* (Vocabularios Indígenas 6). México, D.F.: Instituto Lingüístico de Verano, 1962.

Comunicaciones y Transportes (El Ferrocarril Chihuahua al Pacífico), #16. México, Jan.-Feb., 1962.

Díaz, Carlos Infante. *Luka (Traducción-Adaptación Al Idioma Tarahumar del Evangelio de Lucas)*, 1999.

Díaz Gutiérrez, Victoriano (cronista de la ciudad). *Puerta a la sierra: Recuento histórico de Cuauhtémoc.* Cuauhtémoc: Ediciones Aster, 2002.

Dunne, Peter Masten. *Early Jesuit Missions in Tarahumara.* Berkeley: University of California Press, 1948.

Elbow, Gary S. "Mennonites." *Handbook of Texas.* Texas State Historical Association, 2011.

El Heraldo de Chihuahua. June 27, 1997.

The Engineering and Mining Journal. July 1907.

Fowler, Richard B. *Arthur Stilwell in Mexico.* 1950 (Kansas City Public Library--Missouri Valley Special Collections).

Flynn, Ken. "Unspoiled Beauties of Sierra Madre Spark Mexico's Inaugural Train Trip for Newsmen." *El Paso Herald Post*, Nov. 27, 1961.

Gardner, Erle Stanley. *Neighborhood Frontiers.* Toronto: William Morrow & Co., 1954.

Habermeyer, Christopher Lance. *Gringos' Curve (Pancho Villa's Massacre of American Miners In Mexico, 1916).* El Paso: Book Publishers of El Paso, 2004.

Hilton, Ken (translator). *Riosi Ra'ichara (El Nuevo Testamento de Nuestro Señor Jesucristo en Tarahumara)*. México: La Biblioteca Mexicana del Hogar, A.C., 1972.

INEGI 2010 Census.

Irigoyen, Ulisis. "Topolobampo, Salida al Mar." Report presented to the Sociedád Mexicana de Geografía y Estadística. September 2, 1943.

Jaime M., Carlos. *San Juanito, Crónica de lo que el tiempo dejó*. Chihuahua: Gobierno del Estado, 2001.

Journal of the Southwest. Vol. 54, No. 1, Spring 2012. Edited by Joseph Carleton Wilder. The Southwest Center: University of Arizona, Tucson.

Katz, Friedrich. *The Life and Times of Pancho Villa*. Stanford: Stanford University Press, 1998.

Kerr, John Leeds, with Frank Donovan. *Destination Topolobampo*. San Marino: Golden West Books, 1968.

Mares Trías, Albino and Don Burgess McGuire. "Hit and Run." *Natural History*, Sept. 1999.

---. *Re'igí ra'chuela (El juego de palillo)*. Chihuahua: Escuela Nacional de Antropología e Historia, Unidad Chihuahua, 1996.

Memoria de la construcción del Ferrocarril Chihuahua al Pacífico. Secretaría de Obras Públicas, 1963.

Merrill, William. *Raramuri Souls, Knowledge and Social Process in Northern Mexico*. Washington, D.C.: Smithsonian Press, 1988.

---. "La identidad ralámuli: una perspectiva histórica." In Claudia Molinari and Eugeni Porras, eds. *La identidad y los pueblos étnicos en la Sierra Tarahumara*, pp.71-103. Ciudad México: Instituto Nacional de Antropología e Historia and Congreso del Gobierno del Estado de Chihuahua, 2001.

Mining and Scientific Press. Nov. 4, 1911.

Parks, Walter P. *The Miracle of Mata Ortiz*. Riverside, CA: Coulter Press, 1993.

Pennington, Campbell W. *The Tarahumara of Mexico (Their Environment and Material Culture)*. Salt Lake City: University of Utah Press, 1963.

Pérez de Rivas, Andrés. *My life Among the Savage Nations of New Spain* (Thomas A. Robertson, translator). Los Angeles: The Ward Ritchie Press, 1968.

Pérez Elías, Antonio. "Ferrocarriles." *Enciclopedia de México*. Segunda Edición 1977. Tomo IV.

Pletcher, David M. *Rails, Mines, and Progress: Seven American Promoters in Mexico, 1867-1911*. Port Washington, NY: Kennikat Press, 1958.

Rebo Studios. *Train Ride to the Sky*. Television Documentary. 1994.

Robertson, Thomas A. *Utopia del Sudoeste (Una Colonia Americana en México)*. Los Angeles: The Ward Ritchie Press, 1964.

Rodríguez, Luís González. *Crónicas de la Sierra Tarahumara*. México: Secretaría de Educación Pública, 1948.

---. *Crónicas de la Sierra Tarahumara.* Secretaría de Educación Pública, 1987.
---. *El Noroeste Novohispano en la Época Colonial, and Crónicas de la Sierra Tarahumara.* México: Instituto de Investigaciones Antropológicas UNAM, 1993.
Sariego Rodríguez, Juan Luís. *El Indigenismo en la Tarahumara.* México: INI-INAH, 2002.
Schalkwijk, Bob, Luís González Rodríguez, and Don Burgess. *Tarahumara.* México, D.F.: Chrysler de México, S.A., 1985.
Schmidt, Robert. *Map of the Sierra Tarahumara.* El Paso: International Map Co. UTEP (both earlier and latest editions).
---. *Mexico's Sierra Madre Occidental: The Geography.* Forthcoming.
Schmiedehaus, Walter. *Ein feste Burg ist unser Gott (Der Wanderweg eines christlichen Siedlervolkes).* Cuauhtémoc, Chih., México: Druck G.J. Rempel, Blumenort, 1948.
Secretaría de Comunicaciones y Transportes y Ferronales. *Caminos de Hierro.* 1996.
Sheppard, Grant. *The Silver Magnet: Fifty Years in a Mexican Silver Mine.* New York: E.P. Dutton, 1938.
Sonnichsen, C.L. *Colonel Greene and the Copper Skyrocket.* Tucson: Univ. of Arizona Press, 1983.
Stilwell, Arthur Edward. *Cannibals of Finance, Fifteen Year's Contest with the Money Trust.* Chicago: The Farnum Publishing Co., 1912.
The Texas, Topolobampo & Pacific Railroad and Telegraph Company: Reports of Geo. W. Simmons, Jr.., Dr. B. R. Carman, and John E. Price, Esq., Upon the Route of a Railroad from Topolobampo Bay, on the Gulf of California, to Piedras Negras, on the Rio Grande. Boston: Press of Rockwell and Churchill, 1881.
Tompkins, Frank. *Chasing Villa (The Last Campaign of the US Calvary).* Silver City: High-Lonesome Books, 1996.
Torres Gándara, José Luis. *Reseña Histórica de La Junta y del Ferroarril Chihuahua al Pacífico.* Chihuahua: Ediciones del Azar A.C., 2007.
Torres González, Rodolfo. *Un sueño una realidad...Ferrocarril Chihuahua al Pacífico.* Litográfica Voz, 2008.
Velasco Gil, Carlos Mario. *La Conquista del Valle del Río Fuerte.* México, 1957. Rpt. Siglo XXI, 2003.
Wampler, Joseph. *Mexico's "Grand Canyon".* Berkeley: Joseph Wampler, 1978.
Waters, L.L. *Steel Trails to Santa Fe.* Lawrence: The University of Kansas Press, 1950.
Yáñez Bordier, Ing. José Eloy. "Apuntes Históricos de las rutas del Ferrocarril Chihuahua al Pacífico." *Ingeniería Civil,* March 9, 1983.
Zingg, Robert. *Behind the Mexican Mountains.* Austin: University of Texas Press, 2001.

Index to Photographs and Maps

Places:

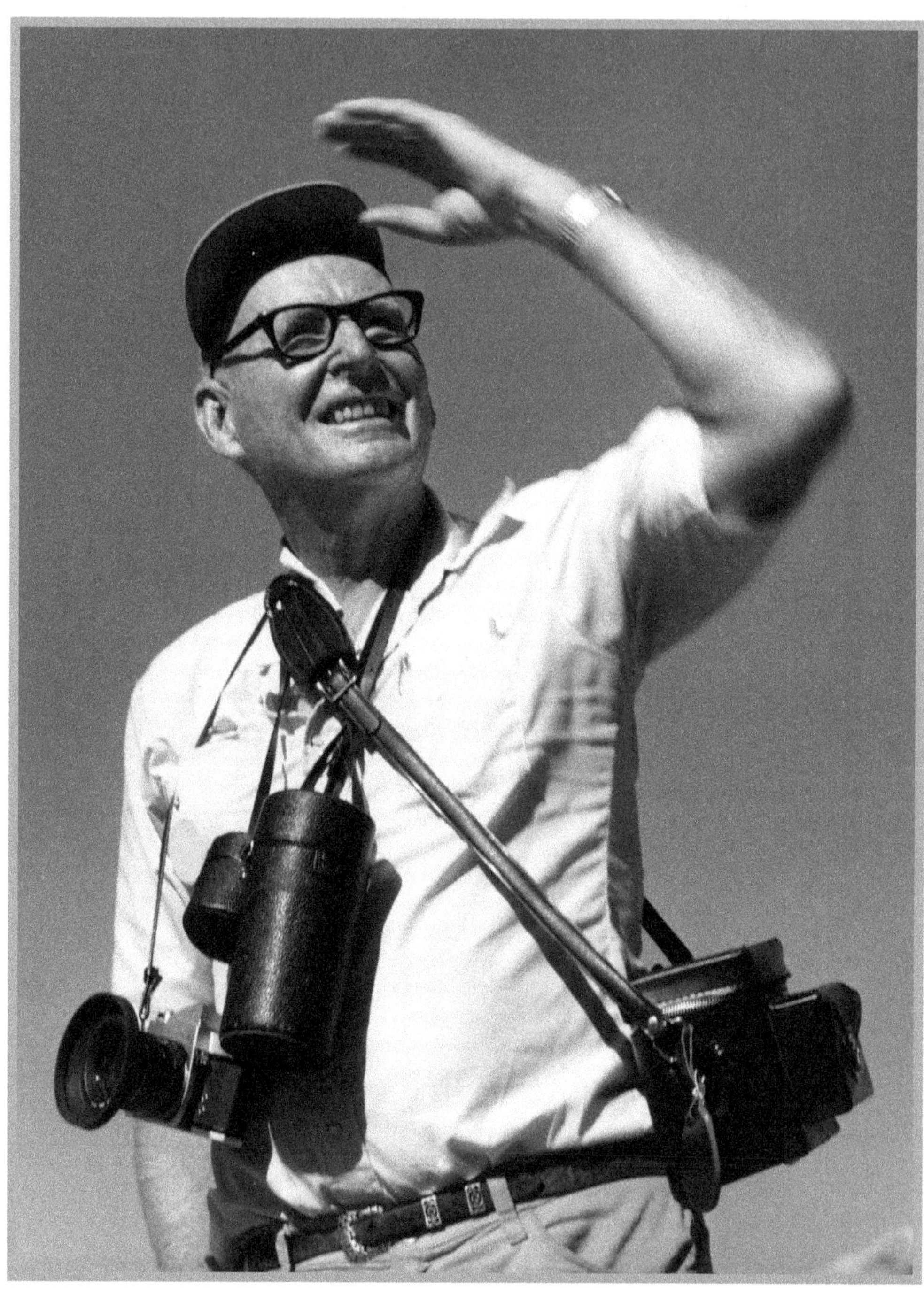

9.4. Glenn leading a photography field trip near Safford, Arizona (Photo taken by an inmate from the Arizona State Prison Complex, where Glenn taught photography for fifteen years, c.1975).

Glenn Burgess:

Glenn (1905-1995) worked as a photo-journalist for the *El Paso Times* and the *Fort Worth Star-Telegram*. At the time, he also taught journalism and photography at Sul Ross College in Alpine, Texas.

Don Burgess:

Don (1939-present), his son, has written numerous books for and about the Tarahumara of Chihuahua, Mexico. He is a linguist/translator and learned photography from his father.

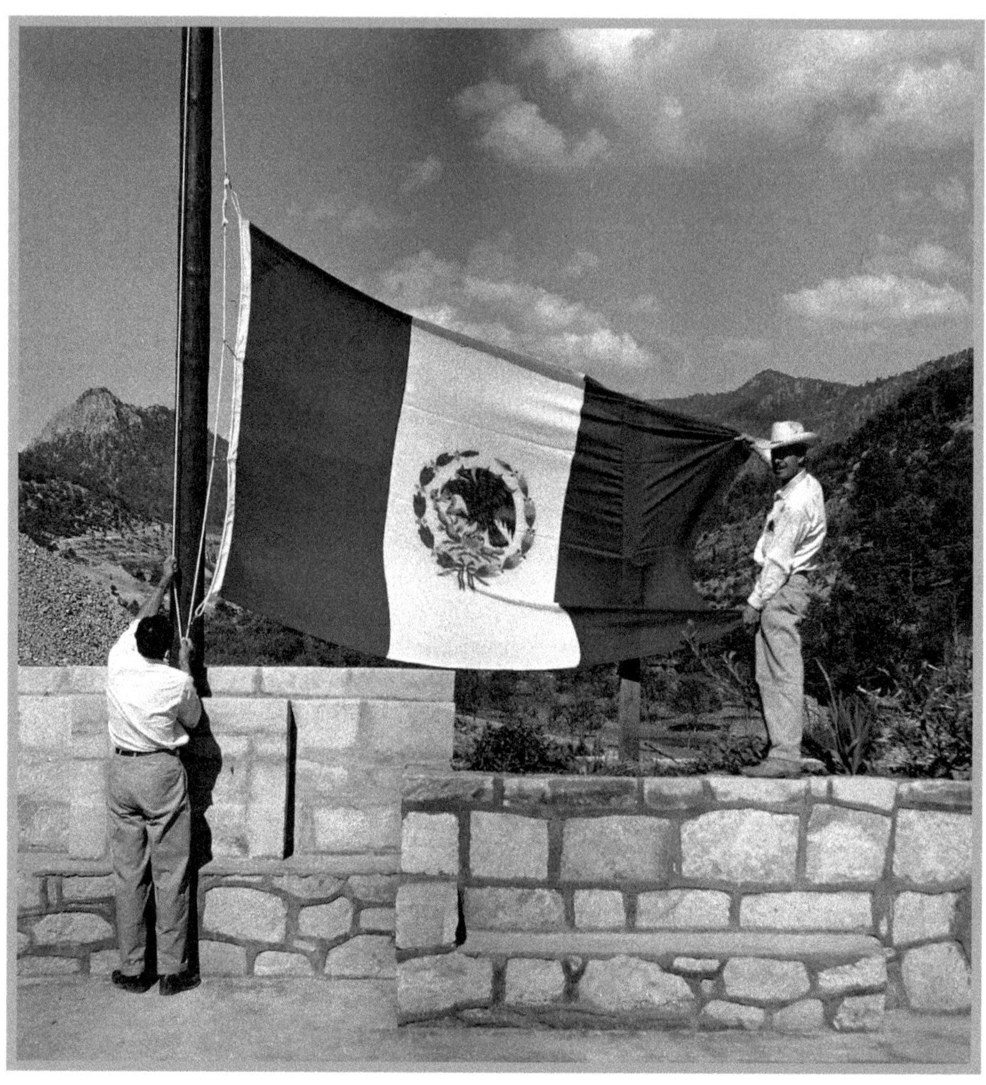

9.5. Ing. Duarte, on left, lowering the Mexican flag at the SOP headquarters at Cuiteco (photo D.B. 1959).

www.ingramcontent.com/pod-product-compliance
Lightning Source LLC
LaVergne TN
LVHW070459120826
845154LV00019BA/88